AF576067

Assessing the Environmental Adaptation of Wildlife and Production Animals

Assessing the Environmental Adaptation of Wildlife and Production Animals: Applications of Physiological Indices and Welfare Assessment Tools

Editor

Edward Narayan

MDPI • Basel • Beijing • Wuhan • Barcelona • Belgrade • Manchester • Tokyo • Cluj • Tianjin

Editor
Edward Narayan
The University of Queensland
Australia

Editorial Office
MDPI
St. Alban-Anlage 66
4052 Basel, Switzerland

This is a reprint of articles from the Special Issue published online in the open access journal *Animals* (ISSN 2076-2615) (available at: https://www.mdpi.com/journal/animals/special_issues/environmental_adaptation).

For citation purposes, cite each article independently as indicated on the article page online and as indicated below:

LastName, A.A.; LastName, B.B.; LastName, C.C. Article Title. *Journal Name* **Year**, *Volume Number*, Page Range.

ISBN 978-3-0365-0142-0 (Hbk)
ISBN 978-3-0365-0143-7 (PDF)

Contents

About the Editor

Edward Narayan (Senior Lecturer in Animal Science). Dr. Edward Narayan graduated with Ph.D. degree in Biology from the University of the South Pacific and pioneered non-invasive reproductive and stress endocrinology tools for amphibians—the novel development and validation of non-invasive enzyme immunoassays for the evaluation of reproductive hormonal cycle and stress hormone responses to environmental stressors. Dr. Narayan leads the Stress Lab (Comparative Physiology and Endocrinology) in the School of Agriculture and Food Sciences (SAFS) at the University of Queensland, a dynamic career research platform that is based on the thematic areas of comparative vertebrate physiology, stress endocrinology, reproductive endocrinology, animal health and welfare, and conservation biology. Edward has supervised 50 research students and published over 70 peer-reviewed scientific research articles.

Editorial

Introduction to the Special Issue: Assessing the Environmental Adaptation of Wildlife and Production Animals: Applications of Physiological Indices and Welfare Assessment Tools

Edward Narayan [1,2]

[1] School of Agriculture and Food Sciences, Faculty of Science, The University of Queensland, St Lucia, QLD 4072, Australia; e.narayan@uq.edu.au; Tel.: +61-7-5460-1693

[2] Queensland Alliance for Agriculture and Food Innovation, The University of Queensland, St Lucia, QLD 4072, Australia

Received: 30 November 2020; Accepted: 2 December 2020; Published: 3 December 2020

Wild animals under human care as well as domesticated farm production animals are often exposed to environmental changes (e.g., capture and transportation). Short-term or acute changes in physiological indices (e.g., heart rate, respiration, body temperatures, immune cells and stress hormonal biomarkers) provide crucial information regarding the responses of animals to novel environments, and they could provide crucial determining factors for evaluating the long-term health and welfare of animals. The goal of this special issue is to provide examples of new research and techniques that can be used to monitor short- and long-term environmental adaptation of animals under human care. Examples of research include applications of physiological indices and welfare assessment methods (e.g., morphological and morphometric data, behavioural assessments, thermal profiles and physiological markers) in any wildlife or production animal (e.g., rescued and rehabilitating animals, pets, competition animals, farm animals and zoo animals), in response to environmental and management-related factors.

This book is a reprint of the papers published in the Special Issue: "Assessing the Environmental Adaptation of Wildlife and Production Animals: Applications of Physiological Indices and Welfare Assessment Tools".

Chapter 1—This research was based on a retrospective analysis of clinical data and characterises this based on categories of stress experienced by avian wildlife patients. It demonstrated the factors associated with urbanization, which exposes avian wildlife to an array of environmental stressors that result in clinical admission and hospitalisation. It showed that the most common outcome of avian patients that suffered from vehicle-related injuries or other impact injuries was euthanasia. Immobility and abnormal behaviour were the most commonly occurring primary stressors of avian patients. Finally, trauma and fractures were the most common occurring secondary stressors in avian patients. The most common outcome of all these stressors was euthanasia. This study also provided a categorisation system for the stressors (preliminary, primary and secondary) that may be used to monitor the stress categories of wildlife patients and gain a deeper understanding of the complex notion of stress.

Chapter 2—Artificial insemination programs are used to improve reproductive output in livestock animals. This study showed the possible influence of ejaculation collection in breeding boars on their oxytocin profiles. Using saliva collection, the research measured total (protein bound) and free oxytocin in pigs. Research showed that ejaculation influences the salivary oxytocin concentrations in breeding boars, although this influence varies according to age, libido and breed.

Chapter 3—Pharmacological and biological validation are important for establishing minimally invasive stress hormone evaluation in animals. Here, researchers were able to validate two faecal glucocorticoid enzyme immunoassays (faecal corticosterone metabolites-FCMs, EIAs; a corticosterone EIA, and a group-specific 5α-pregnane-3β,11β,21-triol-20-one EIA) in deermice (*Peromyscus maniculatus*) by challenging individuals with dexamethasone and adrenocorticotropic hormone (ACTH). Researchers discuss the need for careful physiological/biological validation of assays prior to applications in animal studies.

Chapter 4—Veterinary intervention is an important aspect of wildlife rescue and rehabilitation programs. In this research case study on the roe deer (*Capreolus capreolus*) from Italy, researchers identified a potential biomarker (whole blood lactate concentrations) as an early indicator of lactataemia status to predict the outcome of clinical intervention. Biomarkers such as these should be used in conjunction with clinical veterinary intervention to provide appropriate care and outcome decisions for rescued wildlife.

Chapter 5—Quantification of acute stress is an important component of wildlife management and care in zoos. In this research, the investigators compared the glucocorticoid concentrations in response to various types of potential stressors present during the standard operation of a temporary housing facility between three species, namely, ring-tailed lemurs, collared brown lemurs and white-headed lemurs. Researchers employed polyclonal antibodies directed against the metabolite 11-oxo-etiocholanolone I. Researchers found some species-related differences in the physiological stress responses of the lemurs, however, the general patterns across treatments were similar, but individual reproductive status may also influence the stress responses in grouped housing situation.

Chapter 6—Wildlife hunting is an example of intensive human–animal interaction which can generate physiological stress in animals. Red deer (*Cervus elaphus*) is the target of intensive seasonal hunting in the Lousã Mountain region, Portugal. Using a combination of sample types (blood, faeces and hair), researchers in this study quantified glucocorticoid levels in red deer across hunting seasons. The researchers discuss the applications of specific sample types for evaluating acute and chronic stress responses of red deer in the hunting season.

Chapter 7—Tigers (*Panthera tigris*) are an endangered species and it is crucial to obtain vital health indices of tigers for rescue, rehabilitation and captive management programs. In this pilot study, the researchers demonstrated the application of serum protein electrophoresis as a useful tool in monitoring the health of tigers.

Chapter 8—Stress endocrine response can influence immune response in animals. Similarily, immune response biomarkers can be used to index stress responses in animals. For example, salivary immunoglobulin A (sIgA) has been proposed as a potential indicator of welfare for various species, including Asian elephants, and may be related to adrenal cortisol responses. Here, researchers distinguished circadian rhythm effects on sIgA in male and female Asian elephants and compared patterns to those of salivary cortisol, information that could potentially have welfare implications. Researchers discovered a daily quartic pattern of SIgA in elephants, which can be important information for designing future experiments to standardize field data collection.

Chapter 9—Wild mammals can be highly cryptic and there remains substantial knowledge gaps regarding the physiological control of reproduction in species, such as red deer (*Cervus elaphus* L., 1758). Researchers evaluated the concentration of cortisol and progesterone, extracted from blood and hair, in 10 wild and pregnant red deer females. Researchers successfully quantified cortisol and progesterone in hair samples obtained from the hinds of deer sampled during a regional selective hunting plan. It may be possible the deer were actively breeding during the sampled period and the methods could be used to evaluate the reproductive ecology and welfare of deers used for human hunting purposes.

Chapter 10—The current study developed a baseline welfare assessment protocol for captive Punjab urial adapted from the welfare protocol for domestic sheep from the Welfare Quality® project. It was able to apply the protocol for Punjab urial across two facilities and provided recommendations for areas of improvement for captive management and breeding.

Chapter 11—Koalas (*Phascolarctos cinereus*) are Australia's iconic marsupial species and they face heightened threats from anthropogenic induced environmental change. In this study, the researchers validated the application of a thermal imaging technology (FLIR530TM IR thermal imaging camera) to evaluate heat signatures in koalas in a captive zoo housing facility. The study discussed the technical limitations and applications of this method for tracking heat stress in koalas in zoos.

Chapter 12—Wildlife are commonly impacted by parasites from their surroundings and they can also serve as hosts for zoonotic pathogens. Parasites activate immune responses of wildlife which can be quantified using morphological, blood and tissue analysis. Here, the researchers evaluated the relative impact of parasite pressure vs. parasite load on different host species, using bank voles (*Myodes glareolus*) and wood mice (*Apodemus sylvaticus*) as study species. The researchers sampled sub-adult males to quantify their immune function, infestation load for ectoparasites and gastrointestinal parasites, and infection status for vector-borne microparasites. They used regression trees to find out whether variation in immune indices could be explained by among-site differences (parasite pressure), among-individual differences in infestation intensity and infection status (parasite load) or other intrinsic factors. The research outcome showed that both parasite pressure and parasite load influence the immune system of wild rodents.

Chapter 13—It is important to validate minimally invasive stress hormone assays for each species due to potential species-specific differences in metabolism and excretion of steroids. In this study, the researchers physiologically validated a faecal cortisol metabolite (FCM) enzyme-immunoassay for male reindeer. Researchers conducted a physiological validation of an 11-oxoaetiocholanolone enzyme immunoassay (EIA) for measuring faecal cortisol metabolites (FCMs) in male reindeer by the administration of an adrenocorticotrophic hormone. Researchers also identified the faecal samples belonging to individual animals using DNA analysis across time. This study reports a successful validation of a non-invasive technique for measuring stress in reindeer, which can be applied in future studies in the fields of biology, ethology, ecology, animal conservation and welfare.

Chapter 14—Minimally invasive hormone monitoring methods can be used to evaluate the physiological responses of aquatic fish species to habitat quality. Here, the researchers validated mucous cortisol assays in wild freshwater fish (Catalan chub, *Squalius laietanus*) living across a pollution gradient. They compared the mucous cortisol levels with cortisol levels in blood and haematological parameters. The results showed that the variation in cortisol in skin mucus followed a similar pattern of response to that detected by the quantification of cortisol levels in blood and the hematological parameters, such as erythrocytic alterations and neutrophil to lymphocyte ratios. Skin mucus could be potentially used as a biomarker in fish welfare evaluation.

Funding: This research received no external funding.

Conflicts of Interest: The author declares no conflict of interest.

Article

Identifying the Stressors Impacting Rescued Avian Wildlife

Kimberley Janssen [1], Crystal Marsland [1], Michelle Orietta Barreto [2], Renae Charalambous [1,3] and Edward Narayan [1,3,4,*]

1 School of Science and Health, Western Sydney University, Locked Bag 1797, Penrith, NSW 2751, Australia; 17724337@student.westernsydney.edu.au (K.J.); 18660193@student.westernsydney.edu.au (C.M.); r.charalambous@uq.net.au (R.C.)
2 School of Veterinary Sciences, Faculty of Science, University of Queensland, St Lucia, QLD 4072, Australia; m.barreto@uq.net.au
3 School of Agriculture and Food Sciences, Faculty of Science, University of Queensland, St Lucia, QLD 4072, Australia
4 Queensland Alliance for Agriculture and Food Innovation, University of Queensland, St Lucia, QLD 4072, Australia
* Correspondence: e.narayan@uq.edu.au; Tel.: +61-7-5460-1693

Received: 6 July 2020; Accepted: 24 August 2020; Published: 25 August 2020

Simple Summary: Stress evaluation in wildlife is valuable tool for rehabilitation and injury prevention. This pilot study investigated categories of stress in rescued birds. We determined three categories of stressors (preliminary, primary and secondary) using clinical data of rescued birds from Adelaide, South Australia. It was discovered that birds are highly susceptible to impact injuries (e.g., flying into a building window) and vehicle-related injuries as preliminary stressors, which often result in hospitalisation of birds. Immobility and abnormal behaviour represented the most common primary stressor, while the most common secondary stressors included trauma and fracture. Furthermore, the most common outcome in clinics due to exposure of birds to these three stressor categories was euthanasia.

Abstract: Urbanisation exposes avian wildlife to an array of environmental stressors that result in clinical admission and hospitalisation. The aim of this pilot study was to conduct a retrospective analysis of clinical data and characterise this based on categories of stress experienced by avian wildlife patients. The results from this study indicated that impact injuries (n = 33, 25%) and vehicle-related injuries (n = 33, 25%) were the most common occurring preliminary stressors that resulted in the hospitalisation of avian wildlife. The most common outcome of avian patients that suffered from vehicle-related injuries was euthanasia (n = 15, 45%), as was avian patients that suffered from impact injuries (n = 16, 48%). Immobility (n = 105, 61%) and abnormal behaviour (n = 24, 14%) were the most commonly occurring primary stressors of avian patients. Finally, trauma (n = 51, 32%) and fractures (n = 44, 27%) were the most common occurring secondary stressors in avian patients. The most common outcome of all these stressors was euthanasia. This study provided further evidence towards the notion that human- and urbanisation-related stressors are the main causes of hospitalisation of avian wildlife, but also indicated that birds admitted as a result of human-related stressors are more likely to be euthanised than released. This study also provided a categorisation system for the stressors identified in avian wildlife patients (preliminary, primary and secondary) that may be used to monitor the stress categories of wildlife patients and gain a deeper understanding of the complex notion of stress.

Keywords: wildlife; environmental stress; urbanisation; birds

1. Introduction

Clinical treatment for injured avian wildlife is well explored within the literature [1–3], however, there is limited information regarding the long-term impacts that environmental stress has on the recovery of a patient. Environmental stressors are factors within the environment that cause stress to an individual [4]. Examples of environmental stressors include biotic factors, such as limited/reduced food availability, presence of predators, existence of pathogenic organisms, and interactions with conspecifics [4]. Alternatively, abiotic factors exist, such as extreme temperatures, reduced water availability, and the presence of toxicants [4]. Currently, the main limitation in clinical avian care research is that little is known about how environmental stressors affect avian wildlife.

The universal meaning of stress has been difficult to define. Moberg [5] defined stress as 'the biological response elicited when an individual perceives a threat to its homeostasis'. This definition has since been debated particularly due to the word "homeostasis" [6]. Nevertheless, it is generally agreed upon that stress is a biological response, termed as the stress response, that occurs when an animal is presented with an unpleasant stimulus known as a stressor [7,8]. Stress is not inherently harmful; however, ongoing stress has pervasive consequences for the well-being of animals and dictates the long-term survival and quality of life of veterinary patients [9–11]. When animals encounter environmental stressors, the hypothalamic–pituitary–adrenal (HPA) axis is activated, which prepares the body for some form of exertion [12,13]. The hypothalamus then releases a hormone called corticotrophin releasing factor (CRF), which signals the anterior pituitary to release a hormone called adrenocorticotrophic hormone (ACTH) [12,13]. Adrenocorticotrophic hormone circulates in the blood and results in an increased output of glucocorticoids from the adrenal cortices [12,13]. Glucocorticoids act to divert the storage of glucose as glycogen, and to instead mobilise glucose from stored glycogen [12,13]. The most pivotal glucocorticoid within the HPA axis is cortisol, and it works to stimulate gluconeogenesis [12,13]. Gluconeogenesis acts in a way that prepares the animal for a physical challenge by partitioning energy and also acts as a chemical blocker within the negative feedback process [12,13]. Since the HPA axis comes at a cost of diverting energy away from corporal bodily functions, long-term exposure to environmental stressors can reduce growth, reproduction, and immune function in animals [12].

Four categories have been used to quantify stress in fish [9]. These categories include primary stress, secondary stress, and tertiary stress [9]. For the purpose of this study, these categories were adapted to avian patients in clinical care and a fourth category, preliminary stress, was introduced. Preliminary stress refers to the initial causative factor that resulted in a patient requiring any sort of treatment in a clinical setting. A preliminary stressor is anything that can cause any physical or psychological stress to an individual. This may include an animal attack, vehicle collision or heat stress. Primary stress refers to the effect caused by preliminary stress including any physical or behavioural abnormalities [9]. This may include abnormal behaviour, feather damage or bleeding. Secondary stress refers to the diagnosis which resulted in or caused the preliminary stressor [9]. This may include fractures, disease and infection. Tertiary stress refers to a long-term stressor that may impact a patient after the other stressors have been treated [9]. This may include brain damage, permanent body disfigurement and loss of sight or other senses. For example, if a bird flew into a window, it would have experienced a preliminary stressor. If the wings of this bird had begun to bleed, it would have experienced the bleeding as a primary stressor. If this bleeding was due to a broken bone which had punctured the skin, it would have experienced the fracture as a secondary stressor. Finally, if the broken bone had resulted in permanent body disfigurement and an inability to fly properly, it would have had experienced a tertiary stressor. Beyond categorising the complex notion of stress for the purpose of gaining a clearer understanding of this biological phenomenon, this would also help us to minimise the intensity and frequency of stress experienced by animals, which are two very significant characteristics of stress involved in wildlife recovery [10–13]. Therefore, it is integral to quantify the chain of stressors experienced by wildlife in clinical care for the control of these stressors from when the bird is rescued, throughout treatment and after release or rehoming.

Current clinical data surrounding the assessment and management of stress in avian wildlife admitted to clinical care are often difficult to follow. This is due to stress management often requiring invasive methods such as blood collection. However, research using existing records from wildlife hospitals could be used as a tool to better understand avian preservation efforts, particularly of species under conservation [14–19]. Furthermore, these databases can help us understand the impact of human activities on wildlife in a particular geographic location and how this impact varies among different avian species, age and human rural versus urban living environments [19]. Lastly, wildlife records could also illuminate the typical outcome of avian recues i.e., the likelihood of recovery and release versus death, and the circumstances surrounding these outcomes.

The aim of this study was to conduct a retrospective analysis of clinical data and characterise this based on categories of stress experienced by avian wildlife patients admitted to a wildlife clinic. This form of clinical intervention aims to serve as a database for ecological research and urban planning.

2. Materials and Methods

This study was conducted in collaboration with the Adelaide Koala and Wildlife Hospital (AKWH), located in Plympton, South Australia. Clinical data for avian wildlife patients presented to the hospital between 2014 and 2017 were collected on site at the AKWH. The clinical data collected were used to obtain information on the stressors experienced by avian wildlife patients throughout their stay at the AKWH. These data were then systematically collaborated in a Microsoft Excel document and classified according to the patient's age (egg, nestling, juvenile, or adult), species (magpie, lorikeet, ibis, kookaburra etc.), their classification of stress (preliminary, primary, secondary), and finally, the outcome of that diagnosis (euthanasia, care, release etc.). Tertiary stress unfortunately was not able to be investigated to the expected extent and was intended based on the long-term outcome in correlation to the severity of the patient's condition due to lack of clinical records. Therefore, this category was omitted.

The location in which the birds were found was categorised based on a method outlined by Narayan and Vanderneut [20] and criteria provided by the Australian Bureau of Statistics. Locations were provided by suburb and we used Google maps and location demographics to categorise the suburbs as urban, rural or rural–urban. A location was categorised as "urban" if it was densely population and included a population of more than 1000 people. A location was categorised as "rural" if it included was sparsely populated and consisted of mainly open land and contained few buildings. Finally, an area was described as rural–urban if it was situated near or on a fringe between rural and urban areas and if it was populated to a lesser extent than urban areas but more so than rural areas.

An important caveat to note here is that the data provided were not always comprehensive and there were some information gaps. For example, all entries from 2015 were missing and unable to be collected, and some of the provided entries were missing some information, such as the bird's species or location found. For this reason, the data were too unstable to complete statistical analysis beyond the scope of a descriptive analysis. The purpose of this preliminary study, however, was not to analyse the data per year but to create an average to be used for discussion purposes.

3. Results

A total of 178 records pertaining to birds rescued in 2013 ($n = 6$), 2014 ($n = 37$), 2016 ($n = 51$) and 2017 ($n = 84$) were collected (Supplementary Data: Table S1). The majority of birds were rescued from urban areas ($n = 135$), followed by rural ($n = 6$) and rural–urban ($n = 6$) areas. Note that 31 birds were missing location information. The total number of records was comprised of 25 different types of bird. Of these, lorikeets were the most commonly rescued ($n = 46$), followed by magpies ($n = 25$) and cockatoos ($n = 23$). The most common age group of rescued birds was adult ($n = 143$), followed by juvenile ($n = 20$) and nestling ($n = 15$).

Results from this study show that impact injuries ($n = 33$, 25%) and vehicle-related injuries ($n = 33$, 25%) were the most common occurring preliminary stressors which caused hospitalisation of avian patients (Figure 1). Note that vehicle injuries may be documented as impact injuries if there were no witnesses or evidence that a car was involved upon the admission of a bird. The most common outcome of avian patients that suffered from vehicle injuries was euthanasia (n = 15, 45%) and only 18% (n = 6) were released back into their ecosystem. Likewise, the most common outcome for avian patients that had suffered from impact injuries was also euthanasia (n = 16, 48%). A previous study reported that impact injuries are not typically fatal events for birds [21]. However, our results indicate that only 27% (n = 9) of avian patients admitted due to impact injuries were able to be released back into their ecosystem. The remaining patients had no outcome information, died due to their injuries or were kept in care and no further information was provided.

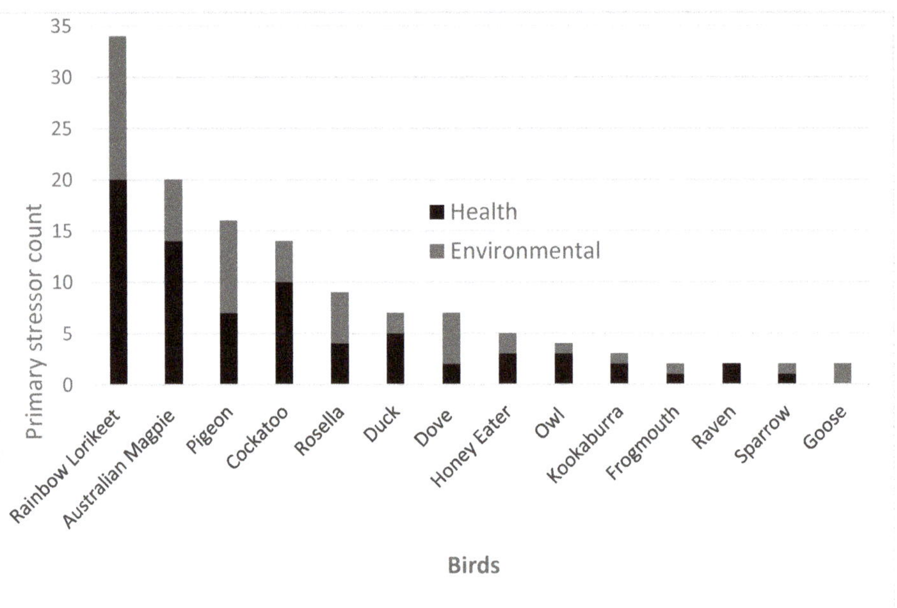

Figure 1. Key preliminary stressors experienced by majority of the avian patients admitted to the Adelaide Koala and Wildlife Hospital in 2013, 2014, 2016 and 2017 (n = 138). Preliminary stressors were pooled into two categories (health or environmental). Health-related preliminary stressors comprised of factors such as impact injury, vehicle trauma, fallen onto ground from substrate, lice presence, severely wet (unable to fly) and genetic issue (not known). Environmental-related preliminary stressors included factors such as animal attack, abnormal behaviour, bullied, rubbish attached, abandoned, ant attack and heat stress. Refer to Supplementary Table S1 to access raw data related to all bird patients, as not all birds have been shown in the above graph where the preliminary stressor count was only one per bird species.

As mentioned in the above caption for Figure 1, abnormal behaviour referred to behaviours that are abnormal for that bird species and age group that were not characterised as immobility. Animal attack (n = 4) also included cat (n = 15) and dog attacks (n = 5). Impact injury included any trauma from events such as flying into a window or building which was not related to vehicle collisions and which did not fit any of the other categories. Abandoned refers to a young bird which was separated from its mother and found alone. Likewise, fallen from nest refers to a young bird who fell but did not land in a pool of water. In contrast, fell in pool could refer to a bird of any age that fell in a pool of water. Bullied refers to birds which had experienced bullying behaviour from other more dominant birds to the point where they required clinical care. Wet included birds which had experienced difficulty flying or locomoting due to wet weather conditions.

Ecological groupings of bird species within preliminary stressors (Figure 1) were as follows: Cockatoo (n = 21) also included yellow-tailed black cockatoo (n = 1), sulphur-crested cockatoo (n = 2) and galah (n = 9). Magpie (n = 19) included one Murray magpie (n = 1). Honey eater (n = 1) also included noisy miner (n = 2), native miner (n = 1) and wattle bird (n = 1). Dove (n = 6) also included one spotted dove (n = 1). Lorikeet (n = 31) also included one musk lorikeet (n = 1). Owl (n = 1) also included boobook owls (n = 3). There were some birds presented to the hospital whose preliminary stressor was unable to be identified (n = 40) and thus were omitted from this figure.

Primary stress refers to the effect caused by preliminary stress including any physical or behavioural abnormalities. Immobility (n = 105, 61%) and abnormal behaviour (n = 24, 14%) were the most common occurring primary stressors (Figure 2). The most common outcome of avian patients that suffered from both immobility and abnormal behaviour was euthanasia at 50% (n = 52) and 38% (n = 9), respectively.

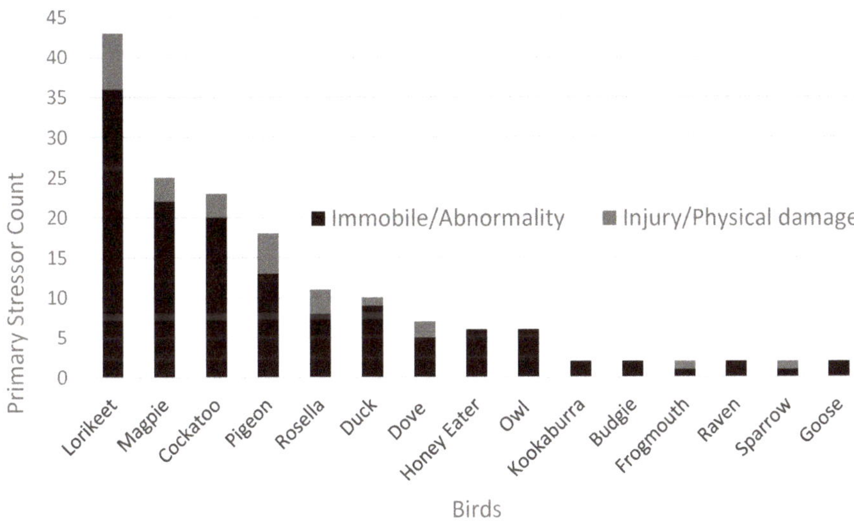

Figure 2. Key primary stressors experienced by the majority of avian patients admitted to the Adelaide Koala and Wildlife Hospital in 2013, 2014, 2016 and 2017 (n = 173). Note that birds presented to the hospital whose primary stressor was unable to be identified (n = 5) or those species with a cumulative count of only one primary stressor were omitted from this figure. Primary stressors were pooled into two categories. Immobile/abnormality-related primary stressors comprised of factors such as physical abnormality, abnormal behaviour and immobile. Injury/physical damage-related primary stressors included factors such as dislocation, oil, damaged feet, diarrhoea, superficial injury, feather damage and bleeding. Refer to Supplementary Table S1 to access raw data related to all bird patients.

Within the primary stressor category (Figure 2), cockatoo (n = 3) also included yellow-tailed black cockatoo (n = 1), sulphur-crested cockatoo (n = 2), galah (n = 15) and corella (n = 2). Magpie (n = 24) also included one Murray magpie (n = 1). Honey eater (n = 2) also included noisy miner (n = 2), native miner (n = 1) and wattle bird (n = 1). Dove (n = 6) also included one spotted dove (n = 1). Lorikeet (n = 42) also included one musk lorikeet (n = 1). Owl (n = 2) also included boobook owl (n = 4).

Secondary stress refers to the diagnosis underlying that of which resulted in or caused the preliminary stressor. Trauma (n = 51, 32%) and fractures (n = 44, 27%) were the most common occurring secondary stressors (Figure 3). The most common outcome of avian patients that suffered from trauma was euthanasia for both trauma (n = 18, 35%) and fractures (n = 25, 57%).

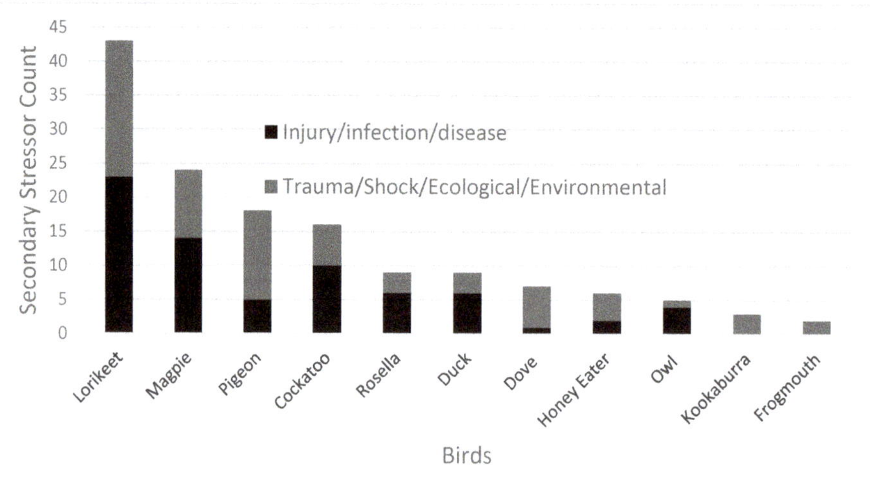

Figure 3. Key secondary stressors experienced by avian patients admitted to the Adelaide Koala and Wildlife Hospital in 2013, 2014, 2016 and 2017 (n = 161). Note that birds presented to the hospital whose secondary stressor was unable to be identified (n = 17) or those species with a cumulative count of only one secondary stressor were omitted from this figure. Secondary stressors were pooled into two categories. Injury/infection/disease-related secondary stressors comprised of factors such as fractures, dislocation, feather damage, broken bone etc. Trauma/shock/ecological/environmental-related primary stressors included factors such as severe dehydration, shock, shot, tissue damage from heat stress etc. Refer to Supplementary Table S1 to access raw data related to all bird patients.

Within the secondary stressor data shown in Figure 3, the ecological groupings were as follows: Cockatoo (n = 2) also included yellow-tailed black cockatoo (n = 1), sulphur-crested cockatoo (n = 2), galah (n = 10) and corella (n = 1). Magpie (n = 23) also included one Murray magpie (n = 1). Honey eater (n = 2) also included noisy miner (n = 2), native miner (n = 1) and wattle bird (n = 1). Dove (n = 6) also included one spotted dove (n = 1). Lorikeet (n = 42) also included one musk lorikeet (n = 1). Owl (n = 2) also included boobook owl (n = 3).

4. Discussion

The results from this study, although restricted to South Australia, can be interpreted on a broader context. Avian ecosystems have undergone profound change due to the increasing threat of urbanisation, which creates disparity in the richness and diversity of the environment [22–24]. Urbanisation challenges avian species by creating threats to their survival through decreased food availability and increased air, light and noise pollution, which results in compromised immune function due to stress [25]. This study allowed these impacts to be quantified by investigating avian patients and, in doing so, identified the risks faced by avian wildlife and the mortality which results from these. Previous studies have attested that anthropomorphically sourced stressors are the main challenges affecting avian wildlife vitality [26–28], and this is consistent with the results of this study. In fact, the preliminary stressors of impact injury, vehicle and rubbish attached, which are all anthropomorphically sourced stressors, accounted for 53% of hospitalisations and 47% of total deaths over a four-year period. These results also indicated that birds which suffer from these stressors are less likely to recover. Furthermore, this study highlighted the fact that these stressors linked to human behaviour are impacting a wide range of avian species. This is important as although the vast majority of the birds identified were not vulnerable or endangered species, the number of wildlife species at risk of endangerment are continuously increasing due to the direct effects of urbanisation and human-related stressors [29]. This indicates that there is much work to be done in order to better preserve Australian avian wildlife.

After implementing a categorical method of assessing stress, veterinary clinics will be able to establish and address different avian stressors, and hopefully implement practices to avoid further hospitalisation and improve mortality rates. When birds suffer with broken bones, their chance of survival is dramatically decreased [14]. This may be due to reasons beyond the severity of the stress experienced by the bird such as the difficulty associated with casting and splinting bones on small mammals [14], and this may also be the reason behind the high euthanasia outcome seen in this study. Furthermore, although it was not able to be investigated in this study, exposure to broken bones may lead to long-term suffering and discomfort from tertiary stressors [12]. These long-term impacts could be due to permanent disfiguration, which can lead to a compromised ability to fly and survive [20]. In our case, the clinic contained no data on the survival rates of an individual or if the patient was ever re-admitted to the same or another clinic. It is recommended that long-term monitoring of once admitted avian patients could be performed after they are released into their natural ecosystems. This would provide statistical data on the tertiary stressors of avian wildlife post-rehabilitation, being that long-term stressors impact a patient after the underlying stressors have been treated. In order to be unnecessarily exhaustive with resources, this long-term monitoring could be used on a selection of patients, such as endangered, native species.

It is important to make note that not all possible clinical cases resulted in rehabilitation, in particular, our results showed that avian patients that were received with bone fracture (n = 25, 57%) were euthanised. The plausible reason for this outcome is due to the length of care and further interaction with human carers that may be required for wild birds that have undergone successful clinical surgery such as bone fracture repair. This could be difficult to manage, especially because wild birds may not be easily desensitised to frequent human exposure, and also because some species of birds (such as rare black cockatoo) may be rehabilitated with a wildlife carer more easily than more commonly occurring species (e.g., magpie). Therefore, a combination of human resource issues as well as infrastructure resource issues may limit the clinical intervention of specific clinical cases such as a bone fracture of wild avian patients, although veterinarians are well trained to perform bone fracture surgeries in avian patients.

Previous studies have highlighted that responses to urbanisation may be species specific. For example, some species of birds disappear completely from an area once the area becomes urbanised, and other species remain and dominate [30]. The categories of stress established in this study can be used to identify species-specific trends of stress, that is, how different species vary in their susceptibility to certain stressors. However, this would involve a level of standardisation for findings to be consistent and reliant. It would be advantageous if everybody (from foster careers to veterinarians) could assess stress using the same formula of evaluation so as to prevent potential disparities. Finally, wildlife recues have been termed potential sentinels of ecosystem health [31]. Although this study was limited in its lack of specifically when it came to diseased patients or those with a microbial or parasitic infection, this is an area for future research which would be particularly useful.

5. Conclusions

This study has contributed fundamental research towards understanding the different categories of stress experienced by avian patients requiring clinical treatment. By organising stressors as preliminary, primary or secondary, this allowed for a clearer understanding of the chain reaction between environmental stress and avian wildlife. Furthermore, this study demonstrated that human- and urbanisation-related stressors were the most common stressors which lead to the hospitalisation and death of birds over a four-year period. In the future, it would be advantageous to monitor tertiary stress in order to allow for an evaluation of birds' wellbeing in the long term. Similarly, these categories of stress could be used to identify species-specific trends and identify which species cope more effectively with urbanisation and which species are more at risk of dying out due to human activity.

6. Patents

Ethics approval and consent to participate: Animal Care and Ethics Committee (ACEC) of the Western Sydney University approved this retrospective study (Approval number: A12145). And we obtained consent from the Adelaide Koala and Wildlife Hospital to publish our research.

Supplementary Materials: The following are available online at http://www.mdpi.com/2076-2615/10/9/1500/s1, Table S1: Raw Data Related to Bird Patients.

Author Contributions: E.N. conceptualised this original research and collaborated with the Adelaide Koala and Wildlife Hospital. E.N. supervised undergraduate field project students K.J. and C.M. for this research project. R.C. is a postgraduate student in the "Stress Lab" (The University of Queensland, Gatton Campus), M.O.B. is a research assistant, and all of the named students provided assistance revising and formatting the manuscript prior to publication. All authors have read and agreed to the published version of the manuscript.

Funding: Research was supported through Western Sydney University and the University of Queensland start-up funds for Edward Narayan.

Acknowledgments: We wish to wholeheartedly thank the veterinary staff, nurses and volunteers of the Adelaide Koala and Wildlife Hospital, especially Phil Hutt and Joanne Sloanne. Thank you to the three anonymous reviewers and Vere Nicolson for valuable feedback during the revision of the manuscript.

Conflicts of Interest: The authors declare no conflict of interest.

Availability of Data and Materials: Clinical data on individual patients are held by the Adelaide Koala and Wildlife Hospital database which is not available to the public. Data relevant to this study are presented in the Results section.

References

1. Guzman, D.S.M. Advances in avian clinical therapeutics. *J. Exot. Pet Med.* **2014**, *23*, 6–20. [CrossRef]
2. González, M.S. Avian articular orthopedics. *Vet. Clin. N. Am. Exot. Anim. Pract.* **2019**, *22*, 239–251. [CrossRef]
3. Brandão, J.; Beaufrère, H. Clinical update and treatment of selected infectious gastrointestinal diseases in avian species. *J. Exot. Pet Med.* **2013**, *22*, 101–117. [CrossRef]
4. Schulte, P.M. What is environmental stress? Insights from fish living in a variable environment. *J. Exp. Biol.* **2014**, *217*, 23–34. [CrossRef]
5. Moberg, G. Biological responses to stress: Implications for animal welfare. In *The Biology of Animal Stress: Basic Principles and Implications for Animal Welfare*; Moberg, G., Mench, J., Eds.; CABI: Wallingford, CT, USA, 2000; pp. 1–22.
6. Romero, L.M.; Dickens, M.J.; Cyr, N.E. The reactive scope model—A new model integrating homeostasis, allostasis, and stress. *Horm. Behav.* **2009**, *55*, 375–389. [CrossRef]
7. Yaribeygi, H.; Panahi, Y.; Sahraei, H.; Johnston, T.P.; Sahebkar, A. The impact of stress on body function: A review. *EXCLI J.* **2017**, *16*, 1057–1072.
8. Fink, G. Stress, definitions, mechanisms, and effects outlined. In *Stress: Concepts, Cognition, Emotion, and Behavior*; Elsevier: Amsterdam, The Netherlands, 2016; pp. 3–11.
9. Schreck, C.B.; Tort, L.; Farrell, A.P.; Brauner, C.J. *Biology of Stress in Fish*; Academic Press: London, UK, 2016.
10. McCormick, G.L.; Shea, K.; Langkilde, T. How do duration, frequency, and intensity of exogenous CORT elevation affect immune outcomes of stress? *Gen. Comp. Endocrinol.* **2015**, *222*, 81–87. [CrossRef]
11. Hing, S.; Narayan, E.J.; Thompson, R.C.A.; Godfrey, S.S. The relationship between physiological stress and wildlife disease: Consequences for health and conservation. *Wildl. Res.* **2016**, *43*, 51–60. [CrossRef]
12. Blas, J. *Stress in Birds*, 6th ed.; Elsevier: Amsterdam, The Netherlands, 2015.

13. Narayan, E.J.; Williams, M. Understanding the dynamics of physiological impacts of environmental stressors on Australian marsupials, focus on the koala (Phascolarctos cinereus). *BMC Zool.* **2016**, *1*, 1–13. [CrossRef]
14. Cousquer, G. First aid in emergency care. *Avian Pract.* **2005**, *27*, 190–203.
15. Kirkwood, J.; Best, R. Treatment and rehabilitation of wildlife casualties: Legal and ethical aspects. *Practice* **1998**, *20*, 214–216. [CrossRef]
16. Mullineaux, E. Veterinary treatment and rehabilitation of indigenous wildlife. *J. Small Anim. Pract.* **2014**, *55*, 293–300. [CrossRef] [PubMed]
17. Cannon, B. Organization for physiological homeostasis. *Physiol. Rev.* **1929**, *9*, 399–431. [CrossRef]
18. Kuenzel, W.J.; Kang, S.W.; Jurkevich, A. Neuroendocrine regulation of stress in birds with an emphasis on vasotocin receptors (VTRs). *Gen. Comp. Endocrinol.* **2013**, *190*, 18–23. [CrossRef]
19. Pyke, G.H.; Szabo, J.K. What can we learn from untapped wildlife rescue databases? The masked lapwing as a case study. *Pac. Conserv. Biol.* **2018**, *24*, 148–156. [CrossRef]
20. Narayan, E.; Vanderneut, T. Physiological stress in rescued wild koalas are influenced by habitat demographics, environmental stressors, and clinical intervention. *Front. Endocrinol.* **2019**, *10*, 1–9. [CrossRef]
21. Hager, S.B.; Cosentino, B.J.; McKay, K.J.; Monson, C.; Zuurdeeg, W.; Blevins, B. Window area and development drive spatial variation in bird-window collisions in an urban landscape. *PLoS ONE* **2013**, *8*, e53371. [CrossRef]
22. Xu, X.; Xie, Y.; Qi, K.; Luo, Z.; Wang, X. Detecting the response of bird communities and biodiversity to habitat loss and fragmentation due to urbanization. *Sci. Total Environ.* **2018**, *624*, 1561–1576. [CrossRef]
23. Sol, D.; Trisos, C.; Múrria, C.; Jeliazkov, A.; González-Lagos, C.; Pigot, A.L.; Ricotta, C.; Swan, C.M.; Tobias, J.A.; Pavoine, S.; et al. The worldwide impact of urbanisation on avian functional diversity. *Ecol. Lett.* **2020**, *23*, 962–972. [CrossRef]
24. Sol, D.; Bartomeus, I.; González-Lagos, C.; Pavoine, S. Urbanisation and the loss of phylogenetic diversity in birds. *Ecol. Lett.* **2017**, *20*, 721–729. [CrossRef]
25. Watson, H.; Videvall, E.; Andersson, M.N.; Isaksson, C. Transcriptome analysis of a wild bird reveals physiological responses to the urban environment. *Sci. Rep.* **2017**, *7*, 1–10. [CrossRef] [PubMed]
26. Palacín, C.; Alonso, J.C.; Martín, C.A.; Alonso, J.A. Changes in bird-migration patterns associated with human-induced mortality. *Conserv. Biol.* **2017**, *31*, 106–115. [CrossRef] [PubMed]
27. Loss, S.R.; Will, T.; Marra, P.P. Direct human-caused mortality of birds: Improving quantification of magnitude and assessment of population impact. *Front. Ecol. Environ.* **2012**, *10*, 357–364. [CrossRef]
28. Loss, S.R.; Will, T.; Marra, P.P. Direct mortality of birds from anthropogenic causes. *Annu. Rev. Ecol. Evol. Syst.* **2015**, *46*, 99–120. [CrossRef]
29. Breed, D.; Meyer, L.C.R.; Steyl, J.C.A.; Goddard, A.; Burroughs, R.; Kohn, T.A. Conserving wildlife in a changing world: Understanding capture myopathy—A malignant outcome of stress during capture and translocation. *Conserv. Physiol.* **2019**, *7*, 1–21. [CrossRef]
30. Isaksson, C. Impact of urbanization on birds. In *Bird Species*; Tietze, D.T., Ed.; Springer: Basel, Switzerland, 2018; pp. 235–257.
31. Trocini, S.; Pacioni, C.; Warren, K.; Butcher, J.; Robertson, I. Wildlife disease passive surveillance: The potential role of wildlife rehabilitation centres. In Proceedings of the National Wildlife Rehabilitation Conference, Canberra, Australia, 22–24 July 2008; pp. 1–5.

Article

Ejaculate Collection Influences the Salivary Oxytocin Concentrations in Breeding Male Pigs

Marina López-Arjona [1], Lorena Padilla [2,3], Jordi Roca [2,3], José Joaquín Cerón [1,*] and Silvia Martínez-Subiela [1]

1 Interdisciplinary Laboratory of Clinical Analysis of the University of Murcia (Interlab-UMU), Regional Campus of International Excellence 'Campus Mare Nostrum', University of Murcia, Campus de Espinardo s/n, 30100 Espinardo, Murcia, Spain; marina.lopez10@um.es (M.L.-A.); silviams@um.es (S.M.-S.)
2 Department of Medicine and Animal Surgery, Veterinary Science, University of Murcia, 30100 Murcia, Spain; lorenaconcepcion.padilla@um.es (L.P.); roca@um.es (J.R.)
3 IMIB-Arrixaca, Regional Campus of International Excellence 'Campus Mare Nostrum', University of Murcia, 30100 Murcia, Spain
* Correspondence: jjceron@um.es; Tel.: +34-868884722

Received: 24 June 2020; Accepted: 16 July 2020; Published: 25 July 2020

Simple Summary: This study aimed to evaluate how the process of ejaculate collection affects oxytocin concentrations in saliva of boars used in artificial insemination. Saliva samples of 33 boars were collected the day before ejaculate collection, during the ejaculation time, and two hours after ejaculate collection. Free oxytocin and oxytocin linked to proteins were quantified in these saliva samples. Oxytocin concentrations during the ejaculation time were higher than the day before with oxytocin linked to proteins showing higher differences. In addition, younger boars, boars with higher libido intensity and boars of the Pietrain breed showed higher values of oxytocin in saliva during ejaculation than the day before. This study demonstrated that ejaculation influences the salivary oxytocin concentrations boars.

Abstract: The objective of the present study was to evaluate the possible changes of oxytocin concentrations in saliva during and after ejaculate collection in breeding boars usually used in artificial insemination programs. Saliva samples of 33 boars were collected the day before ejaculate collection (DB), during the ejaculation time (T0) and two hours after ejaculate collection (T2). Free oxytocin and oxytocin linked to proteins concentrations were measured by two methods previously developed and validated for saliva of pigs. Younger boars, boars with higher libido intensity and boars of the Pietrain breed showed higher values of oxytocin in saliva during ejaculation than the day before. In addition, boars with higher libido showed higher concentrations two hours after ejaculate collection than during the day before. These changes were of higher magnitude and significance when oxytocin linked to proteins was measured. In conclusion, this study demonstrated for the first time that ejaculation influences the salivary oxytocin concentrations in breeding boars, although this influence varies according to age, libido and breed.

Keywords: oxytocin; boar; saliva; ejaculation

1. Introduction

Oxytocin is a neuropeptide hormone that is synthesized in the supraoptic and paraventricular nucleus of the hypothalamus [1]. This hormone has an important role in some physiological functions, such as labor and lactation [2]. In addition, oxytocin inhibits the secretion of glucocorticoids, the hormones associated with anxiety and stress [3]. Moreover, it is increasingly studied in the field of positive emotions and welfare in humans [4,5] and animals [6].

Oxytocin plays an important facilitating role during both male and female reproductive behavior in mammals [7,8]. In male reproduction, oxytocin is thought to be associated with ejaculation by increasing sperm number and contracting smooth muscle of reproductive ducts [9]. Intravenous or intraperitoneal injection of oxytocin contributes to ejaculation in different species, such as rats, rabbits and bulls [10–12]. Similarly, intra-testicular injection of oxytocin increased basal testosterone level in mice [13]. The addition of oxytocin to boar semen artificial insemination (AI) doses improves both sperm transport to the oviduct and conception rates [14–16], but it has no influence on the quality of sperm [15]. In humans, there is an increase in plasma oxytocin levels during, and at 5 min, after ejaculation [17,18].

In pigs, the use of saliva samples for laboratory analysis is a suitable non-invasive and non-stressful alternative to blood. Oxytocin can be measured in pig saliva, it changes throughout lactation period in nursing sows [19] and it is associated to some behaviours [20]. In addition, oxytocin in saliva samples increases in men after sexual self-stimulation [21]. However, to the best of our knowledge, there are no studies about possible changes of oxytocin in saliva related to the ejaculation process in boars. It could be postulated that oxytocin could increase during the ejaculate collection time, possibly due to two main reasons: (1) the role of the oxytocin in the ejaculation, increasing contractility of reproductive smooth muscle [22], and (2) the possible emotional status associated with ejaculation [23].

The purpose of this study was to evaluate possible changes in the salivary oxytocin concentrations during ejaculate collection time in breeding boars. In addition, the influence of breed, age and libido of breeding boars in salivary oxytocin levels was also be evaluated. To accomplish these goals, two new methods previously developed and validated for oxytocin measurement in pig saliva were used, one based in the use of a monoclonal antibody that measures free oxytocin [19], and the other using a polyclonal antibody that measures oxytocin linked to protein [24].

2. Materials and Methods

2.1. Animals and Ejaculate Collection

A total of 33 mature and fertile breeding boars of three different breeds (2 Landrace, 13 Duroc and 18 Pietrain boars) were included in the study. Boars were housed in a farm located in southern Spain. The boars were housed in individual pens in a building with a controlled environment (16 h of light per day and 18–24 °C), with free access to water and fed with commercial feedstuff twice a day. The boars were included in artificial insemination (AI) programs with a regular ejaculate collection of two ejaculates per week. For ejaculate collection, the boars were moved from their pens to another individual pen containing a dummy sow. Once free mounted to the dummy, entire ejaculate was collected using the semi-automatic procedure called Collectis® (IMV technologies, L'Aigle, France).

2.2. Saliva Collection and Oxytocin Measurement

Saliva samples were collected using Salivette tubes (Sarstedt, Aktiengesellschaft & Co. D−51588 Nümbrecht, Germany) that contained a sponge. For sampling, the boars chewed the sponge, which was clipped to a metal rod, until the sponge was moist. The sponges were placed in test Salivette tubes and refrigerated until arrival at the laboratory. The tubes were then centrifuged at 3500 rpm at 4 °C for 10 min. Finally, saliva samples were stored at −80 °C until analysis. The research protocols were approved by the Bioethical Commission of Murcia University according to the European Council Directives regarding the protection of animals used for experimental purposes (approval number, 235/2016; approval date, 25/04/2016).

2.3. Measurement of Salivary Oxytocin Concentration

Saliva samples were thawed at room temperature and oxytocin concentrations were measured using two different methods previously developed and validated for pig saliva [19,24]. One method was a direct competition assay based on AlphaLISA (PerkinElmer Inc., Waltham, MA, USA) technology using monoclonal antibody against oxytocin, while the other consisted of an indirect competition assay based on AlphaLISA technology using polyclonal antibody against oxytocin. Although these assays have been previously described [19,24], in brief, AlphaLISA monoclonal method was performed as follows: 15 μL of sample diluted 1:2 with AlphaLISA buffer and 15 μL of acceptor beads were added to the well and after 90 min, 10 μL of biotinylated oxytocin was added and incubated for 60 min. Then, 10 μL of donor beads was added and after 30 min in the dark, the fluorescence intensity was measured using the Enspire Multimode Plate Reader (PerkinElmer Inc.). In the case of the AlphaLISA polyclonal method, 10 μL of sample diluted 1:2 with AlphaLISA buffer and 10 μL of anti-oxytocin polyclonal antibody were added to the well and after 45 min, 10 μL of protein G acceptor beads was added and incubated for 45 min. Then, 10 μL of biotinylated oxytocin was added and after 45 min, 10 μL of donor beads was added and incubated in the dark for 30 min. Subsequently, the fluorescence intensity was measured using the Enspire Multimode Plate Reader (Perkin Elmer Inc., USA). The accuracy of assays had a correlation coefficient which ranged between 0.96 and 0.99 in the case of the monoclonal method and between 0.98 and 0.99 in the case of the polyclonal method. The intra-assay coefficients of variation were between 5.02–6.07% in the case of the monoclonal method and 4.36–11.37% in the case of the polyclonal method, while the inter-assay coefficients of variation were between 4.60–17.70% in the case of the monoclonal method and 8.52–13.38% in the case of the polyclonal method. The sensitivity of the assays was 112.95 pg/mL and 58.40 pg/mL in the cases of the monoclonal and polyclonal method, respectively.

2.4. Experimental Design

Saliva samples from the 33 boars were collected at three different times in each boar: the day before ejaculate collection (DB); immediately after starting the ejaculation (T0) and two hours after ejaculation (T2). This collection protocol was carried out in February, 2019. Libido was measured according to a three-point scale adapted from that of Kozink et al. [25], namely a value of 1 for boars that showed little interest in the dummy sow and took more than 10 min to mount it; a value of 2 for the boars that did not show much interest but mounted it in less than 10 min; and a value of 3 for the boars that interacted with the dummy and quickly mounted it.

2.5. Statistical Analysis

Medians and 25th 75th percentiles were calculated by use of routine descriptive statistical procedures and computer software (Excel 2016, Microsoft Corporation, Redmond, WA, USA). Statistical analyses were performed using Graph Pad Software Inc (GraphPad Prism, version 5 for Windows, Graph Pad Software Inc, San Diego, CA, USA). The Shapiro–Wilk test was performed to evaluate the data distribution, which did not follow a normal distribution. Therefore, data were log-transformed. One-way repeated measures ANOVA followed by uncorrected Fisher's LSD were used to compare oxytocin values obtained at the different times, namely DB, T0 and T2. Spearman's correlation coefficient was calculated between the analytical parameters with oxytocin concentrations measured with AlphaLISA monoclonal and polyclonal method. The strength of the correlation was assessed by the Rule of Thumb [26], according to which an R value between 0.90 to 1 was considered to have very high correlation, 0.70 to 0.90 high correlation, 0.50 to 0.70 moderate correlation, 0.30 to 0.50 low correlation and less than 0.30 little, if any, correlation. Results were reported as median and 25th–75th percentiles (in text) and line-box plots (in Figures) and a $p < 0.05$ was considered significant.

3. Results

The salivary oxytocin concentrations measured with monoclonal and polyclonal assays showed a low correlation (R = 0.323, $p < 0.05$). As such, the results for each assay are presented separately.

3.1. Oxytocin Concentrations Measured with the Monoclonal Assay

Salivary oxytocin concentrations were significantly higher ($p < 0.05$) at T0 (1077.0 pg/mL; 25–75th percentile: 527.3–2555.0 pg/mL) than at DB time (775.6 pg/mL; 25–75th percentile: 512.8–1494.0 pg/mL) but there were no significant differences ($p > 0.05$) when compared with T2 (802.6 pg/mL; 25–75th percentile: 386.5–1821.0 pg/mL). Interestingly, not all boars showed the same pattern of variation. While some boars showed increased concentrations of oxytocin at T0 compared to DB, other boars showed the opposite behavior (decreased concentrations at T0). Therefore, changes in salivary oxytocin concentrations were evaluated separately in each of the two boar groups. Oxytocin concentrations in the group that showed an increase at T0 ($n = 21$) were higher ($p < 0.01$) at T0 (1767.0 pg/mL; 25–75th percentile: 737.7–3342.0 pg/mL) than at DB (653.4 pg/mL; 25–75th percentile: 413.3–1322.0 pg/mL) and T2 (771.2 pg/mL; 25–75th percentile: 336.5–1821.0 pg/mL). Oxytocin concentrations in the other group, which showed a decrease at T0 ($n = 12$), were lower ($p < 0.05$) at T0 (574.8 pg/mL; 25–75th percentile: 393.2–1069.0 pg/mL) than at DB (1219.0 pg/mL; 25–75th percentile: 550.6–1981.0 pg/mL) and T2 (1022.0 pg/mL; 25–75th percentile: 477.2–1829.0 pg/mL). These results are shown in Figure 1.

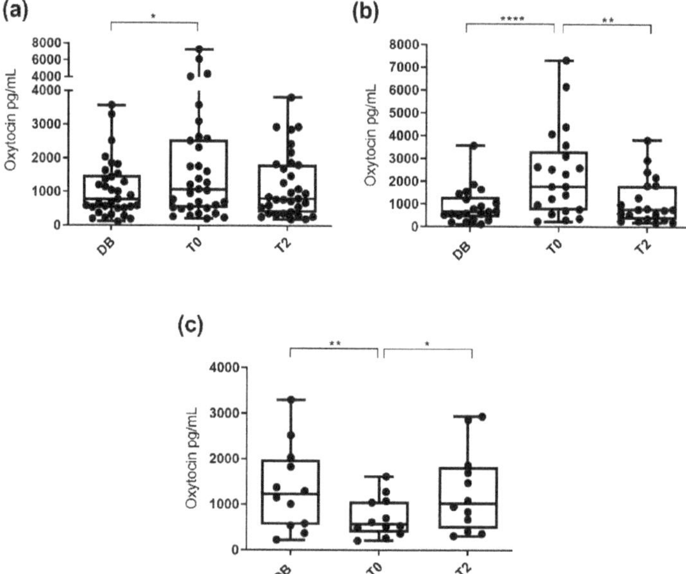

Figure 1. Changes in salivary oxytocin concentrations measured with AlphaLISA monoclonal assay at different times: the day before ejaculate collection (DB), immediately after starting the ejaculation (T0) and two hours after ejaculation (T2). Data for all 33 boars (**a**), data for the 21 boars showing increased oxytocin concentration at ejaculation time (**b**) and data for the 12 boars that did not show increased oxytocin concentrations at ejaculation time (**c**). The plots show medians (line within box), 25th and 75th percentiles (boxes), min and max values (whiskers) and individual data points. Asterisks indicate differences between times (**** $p \leq 0.0001$; ** $p \leq 0.01$; * $p \leq 0.05$).

When pigs were classified by age (boars aged 12 to 24 months, aged 24 to 36 months, and aged more than 36 months), no significant differences in salivary oxytocin between times were seen within each group (Figure 2). However, when the pigs were classified according to their libido, boars with a libido intensity of 3 had higher ($p < 0.05$) oxytocin concentrations at T0 (1250.0 pg/mL; 25–75th percentile: 511.9–2599.0 pg/mL) than DB (834.9 pg/mL; 25–75th percentile: 357.8–1565.0 pg/mL) and did not show significant changes at T2 (749.5 pg/mL; 25–75th percentile: 397.6–1815.0 pg/mL), whereas boars with libido intensity values of 1 and 2 showed similar salivary oxytocin concentrations across the three collection times (Figure 3).

The frequency of individuals having increases in oxytocin concentrations at T0 was different depending on the breed. It was found that around 83.4% of Pietrain boars showed higher oxytocin concentrations at T0 than at DB ($p < 0.01$) and T2 ($p < 0.05$). However, only 46.1% of Duroc boars showed higher oxytocin concentrations at T0 but without differences between DB and T0. The two Landrace boars did not show significant changes at the different times.

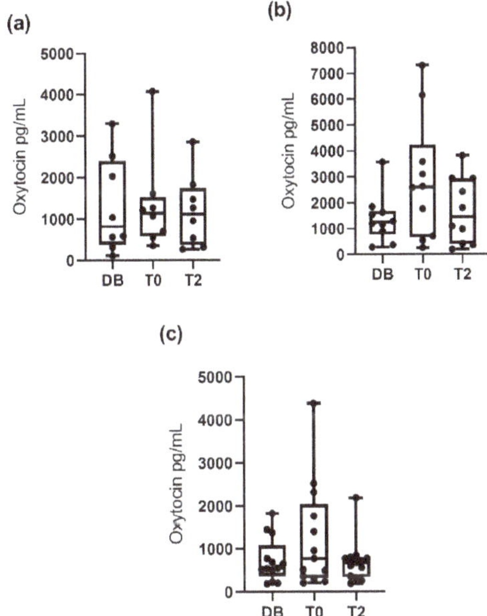

Figure 2. Changes in salivary oxytocin concentrations measured with AlphaLISA monoclonal assay at different times: the day before ejaculate collection (DB), immediately after starting the ejaculation (T0) and two hours after ejaculation (T2). Figure (**a**) shows the data for boars aged 12–24 months and (**b**,**c**) those for boars aged 24–36 and aged more than 36 months, respectively. The plots show medians (line within box), 25th and 75th percentiles (boxes), min and max values (whiskers) and individual data points.

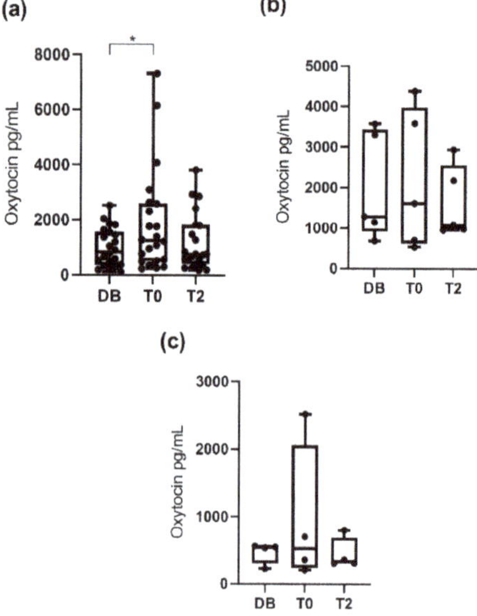

Figure 3. Changes in salivary oxytocin concentrations measured with AlphaLISA monoclonal assay at different times: the day before ejaculate collection (DB), immediately after starting the ejaculation (T0) and two hours after ejaculation (T2). Figure (**a**) shows the data for boars with libido intensity of 3 and (**b**,**c**) those with libido intensity of 2 and 1, respectively. The plots show medians (line within box), 25th and 75th percentiles (boxes), min and max values (whiskers) and individual data points. Asterisk indicate differences between times (* $p \leq 0.05$).

3.2. Oxytocin Concentration Measured with the Polyclonal Assay

Salivary oxytocin concentrations were significantly higher ($p < 0.05$) at T0 (25.8 ng/mL; 25–75th percentile: 7.2–76.7 ng/mL) and T2 (20.3 ng/mL; 25–75th percentile: 11.9–46.5 ng/mL) than at DB time (13.3 ng/mL; 25–75th percentile: 6.1–24.5 ng/mL). Regarding the different pattern of variation, changes in oxytocin concentrations were evaluated separately in each of the two boar groups depending on the increase or decrease at T0. Oxytocin concentrations in the group that showed an increase at T0 ($n = 20$) were higher ($p < 0.05$) at T0 (57.2 ng/mL; 25–75th percentile: 19.9–80.0 ng/mL) and T2 (28.4 ng/mL; 25–75th percentile: 12.4–73.7 ng/mL) than DB (8.7 ng/mL; 25–75th percentile: 5.8–24.8 ng/mL). Oxytocin concentrations in the group that showed a decrease at T0 ($n = 13$) were higher ($p < 0.05$) at DB (13.8 ng/mL; 25–75th percentile: 8.8–23.2 ng/mL) and T2 (15.1 ng/mL; 25–75th percentile: 7.8–24.7 ng/mL) than T0 (6.9 ng/mL; 25–75th percentile: 3.6–11.5 ng/mL). These results are shown in Figure 4.

When pigs were classified by age (boars aged 12 to 24 months, aged 24 to 36 months, and aged more than 36 months), significant differences between times were seen in boars aged 12 to 24 months and in boars aged 24 to 36 months (Figure 5). Oxytocin concentrations observed in boars aged 12 to 24 months were higher ($p < 0.05$) at T0 (67.5 ng/mL; 25–75th percentile: 10.4–108.4 ng/mL) than at DB (13.3 ng/mL; 25–75th percentile: 5.3–39.3 ng/mL). Oxytocin concentrations in boars aged 24 to 36 months were higher ($p < 0.01$) at T2 (32.9 ng/mL; 25–75th percentile: 12.7–79.7 ng/mL) than DB (13.9 ng/mL; 25–75th percentile: 6.1–22.3 ng/mL). Boars aged more than 36 months showed similar salivary oxytocin concentrations at the three collection times. When the pigs were classified according to their libido, boars with a libido intensity of 3 showed higher ($p < 0.05$) oxytocin concentrations at T0 (25.8 ng/mL; 25–75th percentile: 6.9–77.4 ng/mL) and at T2 (24.5 ng/mL; 25–75th percentile: 11.9–56.9 ng/mL) than DB (9.8 ng/mL; 25–75th percentile: 5.9–24.2 ng/mL). Boars with libido intensity 2 and 1 showed similar salivary oxytocin concentrations at the three collection times (Figure 6).

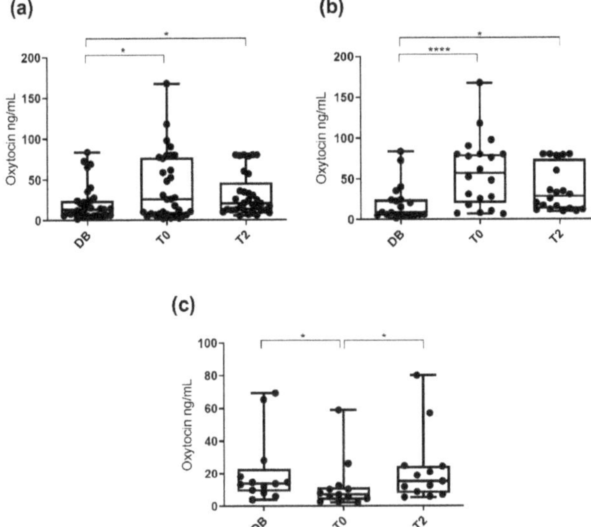

Figure 4. Changes in salivary oxytocin concentrations measured with AlphaLISA polyclonal assay at different times: the day before ejaculate collection (DB), immediately after starting the ejaculation (T0) and two hours after ejaculation (T2). Data for all 33 boars (**a**), data for the 20 boars showing increased oxytocin concentration at ejaculation time (**b**), and data for the 13 boars that did not show increased oxytocin concentrations at ejaculation time (**c**). The plots show medians (line within box), 25th and 75th percentiles (boxes), min and max values (whiskers) and individual data points. Asterisks indicate differences between times (**** $p \leq 0.0001$; * $p \leq 0.05$).

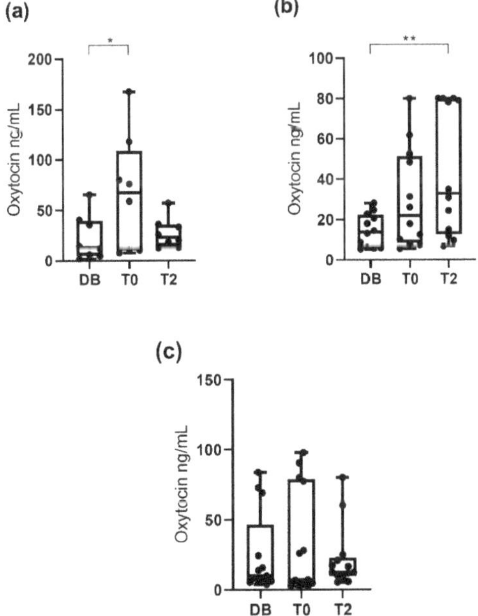

Figure 5. Changes in salivary oxytocin concentrations measured with AlphaLISA polyclonal assay at different times: the day before ejaculate collection (DB), immediately after starting the ejaculation (T0),

and two hours after ejaculation (T2). Figure (**a**) shows the data for boars aged 12–24 months and (**b**) and (**c**) those for boars aged 24–36 and aged more than 36 months, respectively. The plots show medians (line within box), 25th and 75th percentiles (boxes), min and max values (whiskers) and individual data points. Asterisks indicate differences between times (** $p \leq 0.01$; * $p \leq 0.05$).

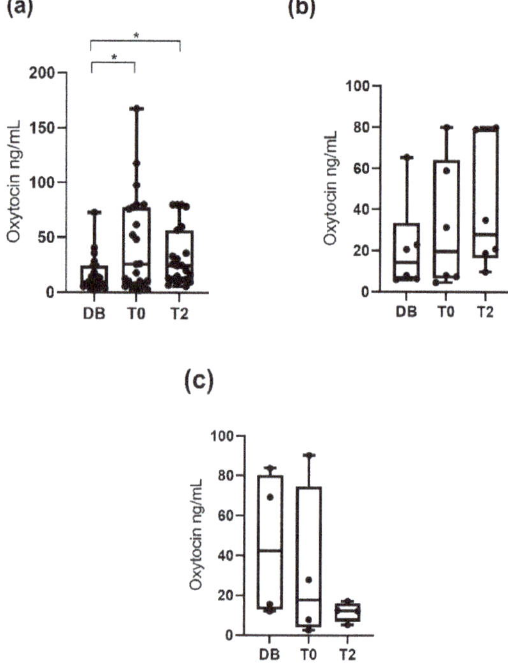

Figure 6. Changes in salivary oxytocin concentrations measured with AlphaLISA polyclonal method at different times: the day before ejaculate collection (DB), immediately after starting the ejaculation (T0) and two hours after ejaculation (T2). Figure (**a**) shows the data for boars with libido intensity of 3 and (**b**,**c**) those with libido intensity of 2 and 1, respectively. The plots show medians (line within box), 25th and 75th percentiles (boxes), min and max values (whiskers) and individual data points. Asterisk indicate differences between times (* $p \leq 0.05$).

The frequency of individuals having increased oxytocin concentrations at T0 was different depending on the breed. We found that 72.2% of Pietrain boars showed higher oxytocin concentrations at T0 than at DB ($p < 0.01$). However, 46.1% of Duroc boars showed an increase at T0 but without differences between DB and T0, although oxytocin concentrations at T2 were higher ($p < 0.05$) than at DB. The two Landrace boars did not show significant changes at the different times.

4. Discussion

This manuscript reports, for the first time, changes in oxytocin concentrations in saliva associated with the ejaculation process in a livestock species. To the best of our knowledge, increases in oxytocin levels after ejaculation have only been reported in cerebrospinal fluid of rats [27], in blood of rabbits [28] and blood and saliva of humans [17,18,21]. All these studies reported an increase of oxytocin levels associated with ejaculation and support our hypothesis that oxytocin would also increase in pig saliva.

Specific studies addressing how oxytocin reaches saliva in pigs have not been reported. However, mammals share essential oxytocin system characteristics, such as production of oxytocin in the hypothalamus and peripheral and central release [29]. Oxytocin is synthesized in the hypothalamic supraoptic and paraventricular nuclei as a neurohormone and is released through the posterior pituitary gland into the bloodstream [30]. In humans, oxytocin reaches the salivary glands via blood circulation through active transport mechanisms [31]. However, the dynamic of oxytocin in both fluids (blood and saliva) could be different, since saliva concentrations of oxytocin do not reflect peripherally circulating oxytocin after exogenous administration in humans [32]. Martin et al. [33] found that oxytocin concentrations in human saliva are more correlated with those in cerebrospinal fluid, and therefore with central oxytocin, than with those of blood plasma.

In this study, two assays were used for measurement of salivary oxytocin, namely monoclonal and polyclonal assays. The results evidenced that the polyclonal assay detects changes of higher magnitude in salivary oxytocin concentrations, especially at 2 h after ejaculate collection. This could be related to the ability of polyclonal assays to detect different metabolites of the oxytocin molecule together with oxytocin bound to other proteins [24] that could remain 2 h after ejaculation. Differences between both assays have also been described in a previous study measuring oxytocin in lactating sows [24]. Together, these results would make polyclonal assays the assay of choice for measuring changes in salivary oxytocin in breeding boars. Accordingly, the discussion will be focused on the results achieved using the polyclonal assay.

The increased salivary oxytocin concentration during ejaculation found in our study reflect its involvement in the ejaculation process [9,34]. Oxytocin contributes to the modulation of sexual behavior [35] and is also involved in the control of male fertility, promoting the transport of sperm through the regulation of basal contractility of smooth muscle in the cauda epididymis [34], as well as stimulation of contraction of the prostate during ejaculation [36]. In addition to its physiological role in the mechanisms related to ejaculation, the increase in oxytocin during ejaculate collection time could also be associated with positive emotions experienced by boars during ejaculation. Oxytocin may represent a link between the social and physical feelings and emotions and sexual behavior [5,37], being released through sensory stimuli, such as touch, warmth and odor [20,38,39]. In addition, oxytocin has become a central component of the mechanisms mediating the well-being and anti-stress effects of positive social interactions in humans [40,41] and animals [6,42].

Interestingly, not all boars showed increased oxytocin concentrations during ejaculation time, as it increased in only 20 of the 33 breeding boars under study. There are several factors that can influence the sexual behavior of boars at ejaculation, such as genetic factors [43]. In this study, more Pietrain than Duroc boars showed increased oxytocin concentrations at T0. The weaker libido of Duroc boars compared with those of others breeds [44,45], as well as the lower total number of sperm and percentage of live sperm in the ejaculates compared with those of Pietrain boars [46], could explain the lower or non-existent increase in salivary oxytocin concentrations at ejaculation time. Furthermore, the age of boars is another influencing factor, as younger boars showed the highest salivary oxytocin concentrations during the ejaculate collection, while the older boars did not show changes between times. We could not find information in the literature to explain differences in salivary oxytocin levels related to age. However, previous studies in boars found that those aged 18 to 24 months had a higher percentage of motile sperm than those aged more than 30 months, which also had higher percentages of sperm abnormalities, such as simple bent tails [47]. In line with this, pregnant women of the first child showed higher serum oxytocin levels compared to pregnant women who already had children [48].

Libido was another factor influencing the salivary oxytocin levels. Boars with high libido showed the highest oxytocin concentrations in saliva at ejaculation time, while boars with moderate or low libido did not show changes in oxytocin concentrations between times. Previous studies in humans found that blood oxytocin was positively related with intensity of sexual behavior and couple interaction [49]. Therefore, it could be postulated that increased libido in boars could be caused by a higher oxytocin synthesis. However, it could also be to the contrary, i.e., increased libido induces increased synthesis of oxytocin. More studies are needed to clarify this issue.

5. Conclusions

In conclusion, this study demonstrated, for the first time, changes in salivary oxytocin concentration in breeding boars during ejaculate collection time. However, the magnitude and significance of changes are different depending on the assay used for its measurement, with the polyclonal assay better suited for detecting changes in oxytocin concentrations than the monoclonal assay. In addition, increases in salivary oxytocin concentrations at ejaculation time in younger boars, in boars showing greater libido intensity, and those of the Pietrain breed, were observed. Overall, this study opens a new line of research about the possible use of salivary oxytocin concentration as potential biomarker of sexual behavior of breeding male pigs. In this context, it would also be interesting to evaluate whether the salivary levels of oxytocin are related to the quantity and quality of semen. This study also suggests the potential use of salivary levels of oxytocin for evaluating the welfare of boars, and for detecting or monitoring problems associated with a low reproductive performance.

Author Contributions: Conceptualization, J.J.C. and J.R.; methodology, M.L.-A. and L.P.; validation, M.L.-A. and S.M.-S.; formal analysis, M.L.-A., J.J.C. and S.M.-S; investigation, M.L.-A., L.P. and S.M.-S; writing—original draft preparation, M.L.-A., J.J.C. and S.M.-S; writing—review and editing, L.P. and J.R.; supervision, J.R., J.J.C. and S.M.-S; funding acquisition, J.J.C. All authors have read and agreed to the published version of the manuscript.

Funding: This research was funded by 'MINECO (Spain), FEDER funds (EU)' (AGL2016−79096-R) and MECD (Spain) FPU16/02170. M.L.-A. was financially supported by 'MECD-FPU' (Madrid, Spain) FPU16/02170.

Acknowledgments: The authors wish to thank the farm workers for their participation in this research.

Conflicts of Interest: The authors declare no conflict of interest.

References

1. Hruby, V.J.; Chow, M.-S. Conformational and structural considerations in oxytocin-receptor binding and biological activity. *Annu. Rev. Pharmacol. Toxicol.* **1990**, *30*, 501–534. [CrossRef] [PubMed]
2. Russell, J.A.; Leng, G. Sex, parturition and motherhood without oxytocin? *J. Endocrinol.* **1998**, *157*, 343–359. [CrossRef] [PubMed]
3. Chiodera, P.; Salvarani, C.; Bacchi-Modena, A.; Spallanzani, R.; Cigarini, C.; Alboni, A.; Gardini, E.; Coiro, V. Relationship between Plasma Profiles of Oxytocin and Adrenocorticotropic Hormone during Suckling or Breast Stimulation in Women. *Horm. Res.* **1991**, *35*, 119–123. [CrossRef] [PubMed]
4. Lopatina, O.L.; Komleva, Y.K.; Gorina, Y.V.; Olovyannikova, R.Y.; Trufanova, L.V.; Hashimoto, T.; Takahashi, T.; Kikuchi, M.; Minabe, Y.; Higashida, H.; et al. Oxytocin and excitation/inhibition balance in social recognition. *Neuropepeptides* **2018**, *72*, 1–11. [CrossRef] [PubMed]
5. Ito, E.; Shima, R.; Yoshioka, T. A novel role of oxytocin: Oxytocin-induced well-being in humans. *Biophys. Physicobiol.* **2019**, *16*, 132–139. [CrossRef]
6. Rault, J.L.; van den Munkhof, M.; Buisman-Pijlman, F.T.A. Oxytocin as an indicator of psychological and social well-being in domesticated animals: A critical review. *Front. Psychol.* **2017**, *8*, 1–10. [CrossRef]
7. Langendijk, P.; Bouwman, E.G.; Schams, D.; Soede, N.M.; Kemp, B. Effects of different sexual stimuli on oxytocin release, uterine activity and receptive behavior in estrous sows. *Theriogenology* **2003**, *59*, 849–861. [CrossRef]
8. Yun, J.; Björkman, S.; Oliviero, C.; Soede, N.; Peltoniemi, O. The effect of farrowing duration and parity on preovulatory follicular size and oxytocin release of sows at subsequent oestrus. *Reprod. Domest. Anim.* **2018**, *53*, 776–783. [CrossRef]
9. Thackare, H.; Nicholson, H.D.; Whittington, K. Oxytocin—Its role in male reproduction and new potential therapeutic uses. *Hum. Reprod. Update* **2006**, *12*, 437–448. [CrossRef]
10. Arletti, R.; Benelli, A.; Bertolini, A. Sexual behavior of aging male rats is stimulated by oxytocin. *Eur. J. Pharmacol.* **1990**, *179*, 377–381. [CrossRef]
11. Melin, P.; Kihlstroem, J.E. Influence of Oxytocin on Sexual Behavior in Male Rabbits. *Endocrinology* **1963**, *73*, 433–435. [CrossRef] [PubMed]

12. Palmer, C.W.; Amundson, S.D.; Brito, L.F.C.; Waldner, C.L.; Barth, A.D. Use of oxytocin and cloprostenol to facilitate semen collection by electroejaculation or transrectal massage in bulls. *Anim. Reprod. Sci.* **2004**, *80*, 213–223. [CrossRef] [PubMed]
13. Gerendai, I.; Csaba, Z.; Csernus, V. Effect of intratesticular administration of somatostatin on testicular function in immature and adult rats. *Life Sci.* **1996**, *59*, 859–866. [CrossRef]
14. Gibson, S.; Tempelman, R.J.; Kirkwood, R.N. Effect of oxytocin-supplemented semen on fertility of sows bred by intrauterine insemination. *J. Swine Health Prod.* **2004**, *12*, 182–185.
15. Okazaki, T.; Ikoma, E.; Tinen, T.; Akiyoshi, T.; Mori, M.; Teshima, H. Addition of oxytocin to semen extender improves both sperm transport to the oviduct and conception rates in pigs following AI. *Anim. Sci. J.* **2014**, *85*, 8–14. [CrossRef] [PubMed]
16. Ngula, J.; Manjarín, R.; Martínez-Pastor, F.; Alegre, B.; Tejedor, I.; Brown, T.; Piñán, J.; Kirkwood, R.N.; Domínguez, J.C. A novel semen supplement (SuinFort) improves sow fertility after artificial insemination. *Anim. Reprod. Sci.* **2019**, *210*, 106193. [CrossRef] [PubMed]
17. Murphy, M.R.; Seckl, J.R.; Burton, S.; Checkley, S.A.; Lightman, S.L. Changes in oxytocin and vasopressin secretion during sexual activity in men. *J. Clin. Endocrinol. Metab.* **1987**, *65*, 738–741. [CrossRef]
18. Ogawa, S.; Kudo, S.; Kitsunai, Y.; Fukuchi, S. Increase in Oxytocin Secretion at Ejaculation in Male. *Clin. Endocrinol.* **1980**, *13*, 95–97. [CrossRef]
19. López-Arjona, M.; Mateo, S.V.; Manteca, X.; Escribano, D.; Cerón, J.J.; Martínez-Subiela, S. Oxytocin in saliva of pigs: An assay for its measurement and changes after farrowing. *Domest. Anim. Endocrinol.* **2020**, *70*, 106384. [CrossRef]
20. Lürzel, S.; Bückendorf, L.; Waiblinger, S.; Rault, J.L. Salivary oxytocin in pigs, cattle, and goats during positive human-animal interactions. *Psychoneuroendocrinology* **2020**, *115*, 104636. [CrossRef]
21. de Jong, T.R.; Menon, R.; Bludau, A.; Grund, T.; Biermeier, V.; Klampfl, S.M.; Jurek, B.; Bosch, O.J.; Hellhammer, J.; Neumann, I.D. Salivary oxytocin concentrations in response to running, sexual self-stimulation, breastfeeding and the TSST: The Regensburg Oxytocin Challenge (ROC) study. *Psychoneuroendocrinology* **2015**, *62*, 381–388. [CrossRef] [PubMed]
22. Carmichael, M.S.; Humbert, R.; Dixen, J.; Palmisano, G.; Greenleaf, W.; Davidson, J.M. Plasma oxytocin increases in the human sexual response. *J. Clin. Endocrinol. Metab.* **1987**, *64*, 27–31. [CrossRef] [PubMed]
23. Bishop, J.D.; Malven, P.V.; Singleton, W.L.; Weesner, G.D. Hormonal and behavioral correlates of emotional states in sexually trained boars. *J. Anim. Sci.* **1999**, *77*, 3339–3345. [CrossRef] [PubMed]
24. López-Arjona, M.; Mateo, S.V.; Escribano, D.; Tecles, F.; Cerón, J.J.; Martínez-Subiela, S. Effect of reduction and alkylation treatment in three different assays used for the measurement of oxytocin in saliva of pigs. *Domest. Anim. Endocrinol.* **2020**, *74*, 106498. [CrossRef] [PubMed]
25. Kozink, D.M.; Estienne, M.J.; Harper, A.F.; Knight, J.W. The effect of lutalyse on the training of sexually inexperienced boars for semen collection. *Theriogenology* **2002**, *58*, 1039–1045. [CrossRef]
26. Hinkle, D.E.; Wiersma, W.; Jurs, S.G. *Applied Statistics for the Behavioral Sciences*, 5th ed.; Houghton Mifflin, EEUU: Boston, MA, USA, 2003.
27. Hughes, A.M.; Everitt, B.J.; Lightman, S.L.; Todd, K. Oxytocin in the central nervous system and sexual behaviour in male rats. *Brain Res.* **1987**, *414*, 133–137. [CrossRef]
28. Stoneham, M.D.; Everitt, B.J.; Hansen, S.; Lightman, S.L.; Todd, K. Oxytocin and sexual behaviour in the male rat and rabbit. *J. Endocrinol.* **1985**, *107*, 97–106. [CrossRef]
29. Carter, C.S.; Pournajafi-Nazarloo, H.; Kramer, K.M.; Ziegler, T.E.; White-Traut, R.; Bello, D.; Schwertz, D. Oxytocin: Behavioral associations and potential as a salivary biomarker. *Ann. N. Y. Acad. Sci.* **2007**, *1098*, 312–322. [CrossRef]
30. Landgraf, R.; Neumann, I.D. Vasopressin and oxytocin release within the brain: A dynamic concept of multiple and variable modes of neuropeptide communication. *Front. Neuroendocrinol.* **2004**, *25*, 150–176. [CrossRef]
31. Gröschl, M. Current Status of Salivary Hormone Analysis. *Clin. Chem.* **2008**, *54*, 1759–1769. [CrossRef]
32. Quintana, D.S.; Westlye, L.T.; Smerud, K.T.; Mahmoud, R.A.; Andreassen, O.A.; Djupesland, P.G. Saliva oxytocin measures do not reflect peripheral plasma concentrations after intranasal oxytocin administration in men. *Horm. Behav.* **2018**, *102*, 85–92. [CrossRef] [PubMed]

33. Martin, J.; Kagerbauer, S.M.; Gempt, J.; Podtschaske, A.; Hapfelmeier, A.; Schneider, G. Oxytocin levels in saliva correlate better than plasma levels with concentrations in the cerebrospinal fluid of patients in neurocritical care. *J. Neuroendocrinol.* **2018**, *30*, e12596. [CrossRef] [PubMed]
34. Frayne, J.; Nicholson, H.D. Localization of oxytocin receptors in the human and macaque monkey male reproductive tracts: Evidence for a physiological role of oxytocin in the male. *Mol. Hum. Reprod.* **1998**, *4*, 527–532. [CrossRef] [PubMed]
35. Ivell, R.; Balvers, M.; Rust, W.; Bathgate, R.; Einspanier, A. Oxytocin and male reproductive function. *Adv. Exp. Med. Biol.* **1997**, *424*, 253–264. [CrossRef] [PubMed]
36. Robbins, S.L.; Kumar, V. *Basic Pathology*, 4th ed.; W.B Saunders Company: Philadelphia, PA, USA, 1987.
37. Hiller, J. Speculations on the links between feelings, emotions and sexual behaviour: Are vasopressin and oxytocin involved? *Sex. Relatsh. Ther.* **2004**, *19*, 393–412. [CrossRef]
38. Mitsui, S.; Yamamoto, M.; Nagasawa, M.; Mogi, K.; Kikusui, T.; Ohtani, N.; Ohta, M. Urinary oxytocin as a noninvasive biomarker of positive emotion in dogs. *Horm. Behav.* **2011**, *60*, 239–243. [CrossRef]
39. MacLean, E.L.; Gesquiere, L.R.; Gee, N.R.; Levy, K.; Martin, W.L.; Carter, C.S. Effects of affiliative human-animal interaction on dog salivary and plasma oxytocin and vasopressin. *Front. Psychol.* **2017**, *8*, 1–9. [CrossRef]
40. Grewen, K.M.; Davenport, R.E.; Light, K.C. An investigation of plasma and salivary oxytocin responses in breast- and formula-feeding mothers of infants. *Psychophysiology* **2010**, *47*, 625–632. [CrossRef]
41. Seltzer, L.J.; Ziegler, T.E.; Pollak, S.D. Social vocalizations can release oxytocin in humans. *Proc. R. Soc. B Biol. Sci.* **2010**, *277*, 2661–2666. [CrossRef]
42. Rault, J.L. Effects of positive and negative human contacts and intranasal oxytocin on cerebrospinal fluid oxytocin. *Psychoneuroendocrinology* **2016**, *69*, 60–66. [CrossRef]
43. Oh, S.H.; See, M.T.; Long, T.E.; Galvin, J.M. Estimates of genetic correlations between production and semen traits in boar. *Asian Australas. J. Anim. Sci.* **2006**, *19*, 160–164. [CrossRef]
44. Savić, R.; Petrović, M. Variability in ejaculation rate and libido of boars during reproductive exploitation. *S. Afr. J. Anim. Sci.* **2015**, *45*, 355–361. [CrossRef]
45. Savic, R.; Marcos, R.; Petrovic, M.; Radojkovic, D.; Radovic, C.; Gogic, M. Fertility of boars—What is important to know. *Biotechnol. Anim. Husb.* **2017**, *33*, 135–149. [CrossRef]
46. Kondracki, S. Breed differences in semen characteristics of boars used in artificial insemination in Poland. *Pig News Inform.* **2003**, *24*, 119N–122N.
47. Jankeviciute, N.; Zilinskas, H. Influence of some factors on semen quality of different breeds of boars. *Vet. Ir. Zootech.* **2002**, *19*, 2130.
48. Prevost, M.; Zelkowitz, P.; Tulandi, T.; Hayton, B.; Feeley, N.; Carter, C.S.; Joseph, L.; Pournajafi-Nazarloo, H.; Ping, E.Y.; Abenhaim, H.; et al. Oxytocin in pregnancy and the postpartum: Relations to labor and its management. *Fron. Public Health* **2014**, *2*, 1–9. [CrossRef]
49. Behnia, B.; Heinrichs, M.; Bergmann, W.; Jung, S.; Germann, J.; Schedlowski, M.; Hartmann, U.; Kruger, T.H.C. Differential effects of intranasal oxytocin on sexual experiences and partner interactions in couples. *Horm. Behav.* **2014**, *65*, 308–318. [CrossRef]

Article

Assessment of the Stress Response in North American Deermice: Laboratory and Field Validation of Two Enzyme Immunoassays for Fecal Corticosterone Metabolites

Andreas Eleftheriou [1,*], Rupert Palme [2] and Rudy Boonstra [3]

1. Wildlife Biology Program, University of Montana, 32 Campus Drive, FOR 109, Missoula, MT 59812, USA
2. Department of Biomedical Sciences, University of Veterinary Medicine, A-1210 Vienna, Austria; rupert.palme@vetmeduni.ac.at
3. Centre for the Neurobiology of Stress, University of Toronto Scarborough, 1265 Military Trail, Toronto, ON M1C 1A4, Canada; boonstra@utsc.utoronto.ca

* Correspondence: andreas.eleftheriou@umontana.edu

Received: 27 May 2020; Accepted: 26 June 2020; Published: 30 June 2020

Simple Summary: If we want to employ stress physiology in the management and conservation of wildlife populations, we need robust methods to quantify stress physiology in the field. Although this is typically done with blood glucocorticoids (GCs), scientists now increasingly use fecal cortisol/corticosterone metabolites (FCMs), which are metabolized GCs excreted in feces. However, immunoassays to measure FCMs need to be validated for each species. North American deermice (*Peromyscus maniculatus*; hereafter deermice) are commonly used in laboratory and field studies. Although a corticosterone radioimmunoassay (RIA) has been validated for deermice, there are no validated enzyme immunoassays (EIAs), which do not require radioactive materials. Through laboratory and field experiments, we validated two EIAs for measuring FCMs in deermice. Researchers can now use these EIAs to evaluate stress physiology in deermice without the need for radioactive materials.

Abstract: Stress physiology is commonly employed in studies of wildlife ecology and conservation. Accordingly, we need robust and suitable methods to measure stress physiology in the field. Fecal cortisol/corticosterone metabolites (FCMs) are now increasingly being used to non-invasively evaluate adrenocortical activity; a measure of stress physiology. However, immunoassays that measure FCMs must be appropriately validated prior to their use and factors that can influence FCMs, such as trap-induced stress, must be considered. Deermice (*Peromyscus maniculatus*) are widely used in scientific studies so that developing methods that appropriately measure their adrenocortical activity is critical. In the laboratory, we tested the suitability of two enzyme immunoassays (EIAs; a corticosterone EIA, and a group-specific 5α-pregnane-3β,11β,21-triol-20-one EIA) in deermice by challenging individuals with dexamethasone and adrenocorticotropic hormone (ACTH). We found that dexamethasone suppressed FCM levels within ~10 h post injection whereas ACTH increased FCM levels within ~2 h post injection. In the field, we found that FCM levels generally increased with more time in trap confinement when using both EIAs. Although we acknowledge low sample sizes (N = 4), our results validated the two EIAs for use with FCMs from deermice.

Keywords: captivity-induced stress; enzyme immunoassay; fecal glucocorticoid metabolites; physiological stress in rodents

1. Introduction

In recent decades, researchers have started to employ stress physiology more often as a tool to evaluate how natural and anthropogenic stressors can affect survival and reproductive success of wildlife populations. Given the widespread use of stress physiology in managing and conserving wildlife, identifying suitable and robust methods for evaluating stress physiology in every species is of paramount importance. This is critical because anthropogenic stressors can induce chronic stress in wildlife, which can lead to pathological perturbations [1,2].

Adrenocortical activity, a measure of stress physiology, is typically evaluated via blood glucocorticoids (GCs) and more recently via fecal cortisol/corticosterone metabolites (FCMs), which are metabolized GCs excreted in feces [2–4]. Evaluation of stress physiology via FCMs is non-invasive and avoids the acute stress effects of capture, handling, and venipuncture [1]. Although researchers are increasingly using FCMs, there are concerns about the methodology used to measure them, such as lack of validation [4]. Immunoassays are validated when they can detect expected changes in FCM levels [4]. Without validation, inference becomes less robust. Thus, immunoassays need to undergo analytical, physiological, and biological validations before they are used to measure FCMs in field settings [4]. Analytical validation may include intra- and inter-assay coefficients of variation and parallelism tests [4]. Validations also must be performed in every species because suitable immunoassays for measuring FCMs can vary even between closely related species [4–6].

North American deermice (*Peromyscus maniculatus*, hereafter deermice) are widely used in biomedical (e.g., [7]), physiological (e.g., [8]), and ecological (e.g., [9]) research. Hence, it may not be surprising that FCM evaluation in deermice has already been investigated by others where a corticosterone radioimmunoassay (RIA; [10,11]) and two corticosterone enzyme immunoassays (EIAs; [12,13]) have been used. Although the RIA was validated before use, the EIAs were not, which makes their use questionable [4]. In addition, all assays used antibodies that bind to corticosterone, which is the predominant GC in *Peromyscus* (e.g., [14]). However, intact corticosterone is essentially absent from feces so their use may be suboptimal [4]. For example, an EIA that used an antibody, which detects FCMs with a 5α-3β,11β-diol structure, demonstrated improved FCM detection in house mice (*Mus musculus*) compared with corticosterone EIAs [15]. Nevertheless, corticosterone immunoassays may still detect FCMs, albeit to a lesser degree, because of cross-reactivity between the corticosterone antibody and FCMs [4]. Commercial corticosterone immunoassays are also relatively easy to acquire and use, although expensive [4]. However, commercial RIAs, unlike EIAs, may be less appealing because they use radioactive materials and require a licensed laboratory for their use [4,16]. To the best of our knowledge, there have been no studies that compared or validated EIAs for measuring FCMs in deermice.

We can physiologically validate immunoassays by using adrenocorticotropic hormone (ACTH) and dexamethasone (a synthetic steroid), which increase and decrease endogenous GC production, respectively [4]. However, to biologically validate immunoassays, we need to use stressors that are biologically relevant to the species of interest. Although immunoassays can be validated biologically if they can track diurnal rhythm changes in FCMs, this should be done in addition to other biological validations [4], such as live trapping, which can increase FCMs (e.g., [17,18]). Knowing when these trap-induced rises in FCMs manifest in feces is also of practical use because they can artificially increase FCMs and lead to erroneous results about baseline adrenocortical activity [19]. This time delay between blood GCs and the appearance of metabolites in feces is species-specific [20]. Specifically, in deermice, [21] found a delay of 4 h before there were trap-induced effects on FCMs, but as they pointed out, the effects could have appeared sooner because they did not sample during a shorter interval (<4 h).

In this study, we had three objectives. Firstly, in the laboratory, we wanted to validate two different EIAs in measuring FCMs using physiological challenges (i.e., dexamethasone and ACTH injections). The immunoassays were a corticosterone EIA [12] and a 5α-pregnane-3β,11β,21-triol-20-one EIA (hereafter referred to as the group-specific EIA, [15]). Secondly, again in the laboratory, we wanted to use the diurnal rhythm in GC secretion to biologically validate both EIAs. Thirdly, in the field, we wanted to investigate temporal effects of trap confinement on FCMs using both EIAs to provide additional biological validation.

2. Materials and Methods

2.1. Acquisition and Husbandry of Laboratory Deermice

We acquired 4 adult deermice (2 females: 2 males) from McMaster University, Ontario, Canada, and transported them to the animal holding facility at the University of Toronto—Scarborough, ON, Canada, in April 2016. These deermice were F1 generation offspring from wild deermice that were originally captured from Nebraska, USA [8]. All were ear-tagged, weighed, and sexed. All deermice were non-reproductive (males were non-scrotal, and females had non-perforate vaginas). They were individually housed in polypropylene cages (47 cm × 26 cm × 20 cm) that were equipped with a wire bottom and a glass water bottle with a stainless-steel nipple. All cages were mounted within a second same-sized cage that was equipped with a fine metal mesh. The wire bottom allowed feces and urine to fall through the bottom of the first cage. Though urine continued to pass through the fine mesh, the feces did not. This arrangement minimized urine contamination of feces and disturbance to the animals [22]. Male and female deermice were kept separate on two different but opposite racks within the same animal facility to limit exposure to odor from the opposite sex. We provided them with *ad libitum* water, rodent chow (LabDiet, St. Louis, MO, USA) and ample cotton bedding as nesting material. Deermice were housed under a 12:12 h dark-light cycle (lights on at 08:00 h) at room temperature (20 ± 5 °C). Ventilation fans were positioned in the wall at either end of the holding facility and operated in a push-pull method (the fan at one end pulled air out of the facility and the other pushed outside air into the facility). This method changed the air in the room 13 times/h. The direction of air flow was parallel to cage racks, which prevented cross contamination between cages that held males and females [23].

2.2. Fecal Sample Collection

We collected fecal samples by following Table 1 and discarded feces contaminated with urine. Forceps were disinfected between individuals during sample collection. If fecal pellets were in excess for an individual, we subsampled to get a representative pooled sample from all areas where the individual had defecated. We then stored samples at −20 °C until analyses. During the acclimation period, fecal samples were collected every 2 h (during the dark and light cycles) whereas during the challenge experiments, samples were generally collected every 2 h during the dark cycle and every 4 h during the light cycle (Table 1). This sampling change occurred due to personnel constraints.

Table 1. Timeline of treatments and samples used in statistical analyses for evaluation of fecal corticosterone metabolites in deermice.

Date	Treatment [1]	Immunoassay	Sample Collection [2] Schedule (h Post Treatment)
May 2–3	Acclimation	Corticosterone EIA Group-specific EIA	70, 72, 74, 76, 80, 82 70, 72, 74, 76, 78, 80, 82
May 3–4	Adrenal suppression	Corticosterone EIA Group-specific EIA	2, 4, 6, 10, 12 2, 4, 6, 10, 12
May 5–6	Adrenal stimulation	Corticosterone EIA Group-specific EIA	2, 4, 6, 10, 12 2, 4, 6, 10, 12

[1] Injections administered at ~20:00 h, [2] Acclimation lasted for ~92 h whereas adrenal treatments lasted for ~48 h (see text for details).

2.3. Acclimation Period

Fecal samples were collected for ~92 h after the animals were transferred to our facility (Table 1). We assumed that samples collected on the last day of acclimation reflected baseline FCMs given that previous work with wild meadow voles (*Microtus pennsylvanicus*) found that FCMs were the lowest by the end of the third day of captivity [24].

2.4. Dexamethasone Suppression Challenge

To test whether EIAs could detect an expected decrease in FCM levels, we injected all deermice with 2.5 mg/kg dexamethasone sodium phosphate (Vétoquinol, Québec, Canada) diluted in sterile 0.9% saline intraperitoneally at ~20:00 h. In this way, each individual was used as its own control. Because dexamethasone doses have not been reported for deermice, we formulated this dose based on studies with other small rodents (e.g., [25–27]). We started to collect samples at 22:00 h for ~48 h although we were unable to analyze samples 12 h post injection (Table 1). No samples were collected at the time of injection.

2.5. ACTH Stimulation Challenge

To test whether EIAs could detect an expected increase in FCM levels, we injected all deermice with 250 µg/ kg ACTH (Cortrosyn, Amphastar Pharmaceuticals, Inc., Rancho Cucamonga, CA, USA) mixed in 0.9% sterile saline solution intraperitoneally at ~20:00 h. Again, each individual was used as its own control and because ACTH doses have not been reported for deermice, we formulated this dose based on above cited studies. As above, we started to collect samples at 22:00 h for ~48 h although we were unable to analyze samples 12 h post injection (Table 1). No samples were collected at the time of injection.

2.6. Field Validation

We carried out two field studies to assess the temporal effect of trap confinement on FCM levels. In both studies, deermice were captured in individual non-folding Sherman traps (H.B. Sherman Traps, Tallahassee, FL, USA) baited with oats and peanut butter, and provided with cotton bedding. In field study 1 (for group-specific EIA), apple slices were also provided. For field study 1, we trapped 20 adult deermice (7 males, 13 females) near Drummond, MT, USA, in June 2017. Only three were non-reproductive. Once trapped, we confined deermice in a trap for either 0–2, 4–6 or 8–10 h prior to processing. To do this, we set traps around dusk and checked them after 2 h. Trapped deermice were either processed for the 0–2 h treatment or left in the trap to be processed for the 4–6 h and 8–10 h treatments, where they spent an additional 4 h and 8 h in the trap, respectively. Deermice were removed from traps by "emptying" contents into a plastic bag. We then sexed, weighed, ear-tagged, and evaluated them for reproductive status. Age was determined based on mass (<14 g = juvenile, 14–17 g = subadult, >17 g = adult, [28]). Reproductive status was determined by the presence of scrotal testes in males, and presence of a perforate vagina, lactation, or pregnancy in females.

In field study 2 (for corticosterone EIA), seven adult deermice (4 males, 3 females) were trapped near Charlo, MT, USA, in August 2017. Only two were non-reproductive. We checked traps after ~4 h of setup when we ear-tagged and collected feces from deermice. Afterwards, all deermice were returned to their respective clean traps where they spent the night until dawn (an additional ~7 h). At that time, they were sexed, weighed, evaluated for reproductive status, and sampled a second time for feces. Age and reproductive status were determined as above.

In both field studies, we collected feces from restrained animals and/or their trap and released individuals on site after processing. Feces contaminated with urine were not collected. We followed field protocols to avoid accidental hantavirus infection [29]. All fecal samples were stored at −80 °C until analysis. All procedures involving animal use were approved by the University of Toronto (protocol # 20011602) and/or by the University of Montana Institutional Animal Care and Use Committees (protocol #s 024-16ALDECS-042616, 027-16ALDECS-051016, 028-16ALDECS-051016). Field work was approved by Montana Fish, Wildlife and Parks (permit #2017-029-W).

2.7. Processing of Feces and Extraction of FCMs

Laboratory and field study 1 fecal pellets were first oven-dried for 1 h at ~60 °C to heat-inactivate hantavirus (if present) and then lyophilized (Labconco Corp., Kansas City, MO, USA) for at least 15.5 h at Pennsylvania State University, State College, Pennsylvania, USA. Laboratory study fecal samples were not pulverized; they remained in pelleted form. However, in field study 1 we did pulverize, using a mortar and pestle, because there were fewer total samples (20 samples). To extract FCMs, we weighed 0.05 g (±0.005 g) of dried pellets/powder. Then, we added 1 mL of 80% methanol to each sample suspension, vortexed at 1500 RPM for 30 min, and centrifuged at 2500× *g* at 22 °C for 20 min [4,15]. Supernatants were decanted and frozen at −20 °C. The extraction of field study 2 samples was slightly different and performed at the University of Montana, Missoula, MT, USA. Fecal pellets (14 samples total) were heat-inactivated and oven-dried for 2 h at ~63 °C to ensure elimination of water, because a lyophilizer was unavailable. We pulverized dried pellets and weighed out 0.04 g (±0.005 g) of powder. The lower threshold weight was chosen because sample weights were generally lower in this field study. The rest of the extraction procedure remained unchanged.

2.8. Immunoassay Methods

For the analysis of FCMs two different EIAs were used. The immunoassays were a corticosterone EIA (commercial kit #K014-H1 or H5, provided by Arbor Assays, Ann Arbor, MI, USA) [12] and a 5α-pregnane-3β,11β,21-triol-20-one EIA (group-specific EIA, measuring FCMs with a 5α-3β,11β-diol configuration [15]). Details of the EIAs including cross-reactions are given by Arbor Assays and Touma et al. [15], respectively. All fecal extracts were diluted with EIA buffer prior to being analyzed. The dilution factor was determined by running pooled sample extract at different dilutions against the standard curve, to identify the one that resulted in ~50% binding for the corticosterone EIA (Figure 1A) and the group-specific EIA (Figure 1B). Through parallelism tests, we showed that serial dilutions of pooled extract tracked the EIA's standard curve (Figure 1). These findings demonstrated that methanol residue in sample extracts did not interfere with assay performance. Consequently, sample extracts were diluted 1:10 for the corticosterone EIA, and 1:200 for the group-specific EIA. However, extracts from field study 2 analyzed with the corticosterone EIA had to be diluted 1:80 instead of 1:10 (change discussed later).

We followed manufacturer's instructions for the corticosterone EIA. However, when we analyzed fecal samples with this EIA for field study 2, we also used a wavelength of 650 nm (reference wavelength) in addition to 450 nm. We followed the protocol described by Touma et al. [15] for the group-specific EIA. All samples were assayed in duplicate. Intra-assay coefficients of variation (CVs) were calculated by averaging all sample CVs and inter-assay CVs were calculated by averaging CVs of low and high concentration controls for all plates (except for field study 2, we could only calculate average of low concentration controls). Intra-assay and inter-assay CVs for the corticosterone EIA were (n = 9) 5.7% and 12.7% (field study 2: 12.8%) and for the group-specific EIA (n = 6) 8.6% and 8.6%, respectively.

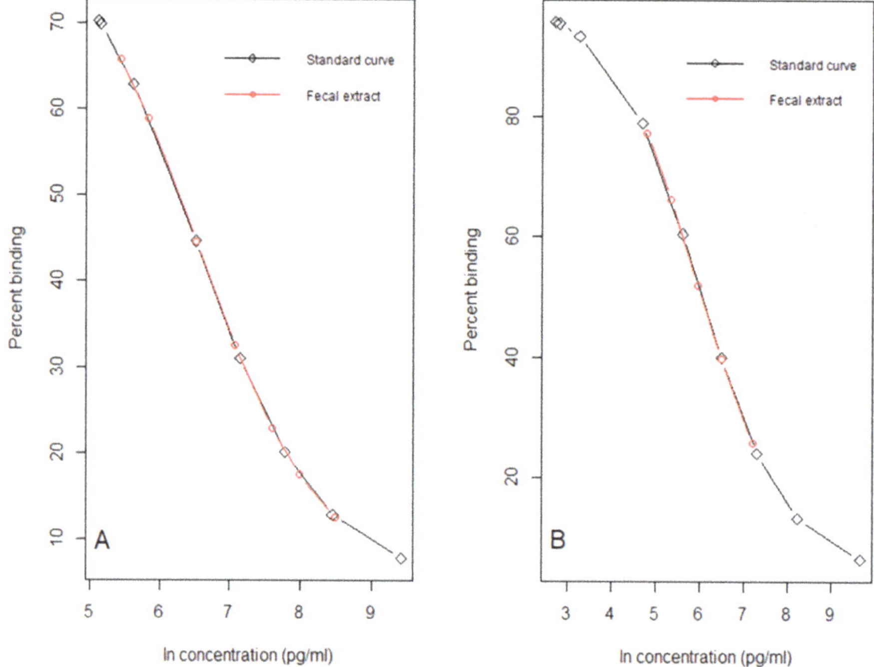

Figure 1. Parallelism curves for pooled fecal extract from deermice measured with (**A**) corticosterone enzyme immunoassay (EIA) and (**B**) group-specific EIA. Standard curves for each EIA are shown in black with each concentration as an open diamond. Parallelism is shown in red with serial dilutions of pooled fecal extract as open red circles.

2.9. Statistical Analyses

All data analyses were done in R [30] within RStudio [31]. We used linear mixed effect models (LMMs) from R packages "lme4" [32] and "lmerTest" [33] to test for diurnal patterns and how each treatment (dexamethasone suppression and ACTH stimulation) affected FCM levels of laboratory-bred deermice compared to baseline. We considered FCMs collected on the last day of acclimation as baseline FCMs. Because deermice were sampled repeatedly, individual identification was included as a random effect, where the model structure was sex + treatment × sampling time, except for when testing for diurnal patterns, which was sex + sampling time. *p*-values were calculated using Satterthwaite's method [33]. We used the R package "emmeans" [34] to perform post hoc pairwise comparisons where *p*-values were adjusted accordingly.

For field study 1, we used a one-way ANOVA with post hoc Tukey's Honest Significant Differences to compare FCM levels across trap confinement treatments (i.e., 0–2, 4–6 and 8–10 h), sex and reproductive status, where the model structure was treatment + sex + reproductive status. For field study 2, we used a LMM to examine the difference between FCMs across two sampling times (i.e., 0–4 h vs. overnight), sex, and reproductive status, where the model structure was treatment + sex + reproductive status. Because each deermouse was sampled twice, individual identification was included as a random effect.

FCM data were ln-transformed prior to all analyses to meet the assumptions of normality and homoscedasticity. Below, we present ln-transformed means with standard errors (ln ng/g of dry feces) where we considered results statistically significant at $\alpha = 0.05$. However, in the figures we present non-transformed data to ease comparisons with other studies.

3. Results

3.1. Acclimation

We did not find changes across sex (p = 0.16) or time (p = 0.10) with the corticosterone EIA (Figure 2A). Similarly, there were no changes across sex (p = 0.60) or time (p = 0.22) with the group-specific EIA (Figure 2B). However, it is noteworthy that the variability in FCMs appeared smaller towards the end of the dark cycle compared to the beginning. FCMs collected during the third day of acclimation were considered as baseline when we evaluated the effects of treatments (dexamethasone and ACTH).

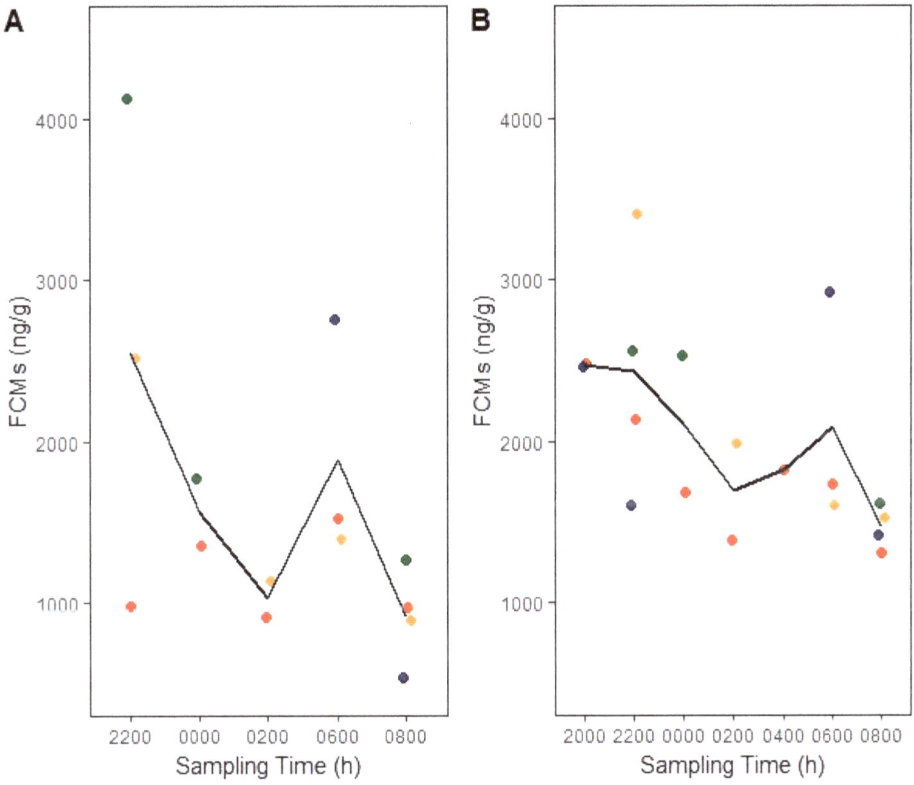

Figure 2. Changes in corticosterone baseline fecal cortisol/corticosterone metabolites (FCMs) from laboratory-bred deermice across the dark cycle measured with (**A**) corticosterone enzyme immunoassay (EIA) and (**B**) group-specific EIA. Lines connect means from each sampling time and circles indicate data points whereby an individual is denoted with a different color. Males are shown in blue and green colors, whereas females are shown in red and tan.

3.2. Dexamethasone Suppression Test

We found a sampling time by treatment effect for the corticosterone EIA ($F_{4,\,18.31}$ = 3.73, p = 0.02; Figure 3A). Deermice had lower FCM levels (n = 4, 6.23 ± 0.26 ln ng/g) ~10 h post injection than baseline (06:00 h, n = 3, 7.54 ± 0.30 ln ng/g, post hoc, $t_{18.32}$ = −3.37, p = 0.003). We also found a sampling time by treatment effect for the group-specific EIA ($F_{4,\,21}$ = 3.39, p = 0.03; Figure 3B). Deermice had lower FCMs (n = 4, 7.03 ± 0.15 ln ng/g) than baseline ~10 h post injection (06:00 h, n = 3, 7.60 ± 0.17 ln ng/g, post hoc, $t_{19.46}$ = −2.47, p = 0.02). Interestingly, deermice showed a marginal increase in FCMs (n

= 4, 8.16 ± 0.15 ln ng/g) ~4 h post injection compared to baseline (n = 2, 7.63 ± 0.22 ln ng/g, $t_{20.42}$ = 1.96, p = 0.06). We note that baseline FCM levels at 06:00 h represent a 2 h interval (04:00–06:00 h) whereas treatment FCM levels at the same time represent a 4 h interval (02:00–06:00 h; Table 1).

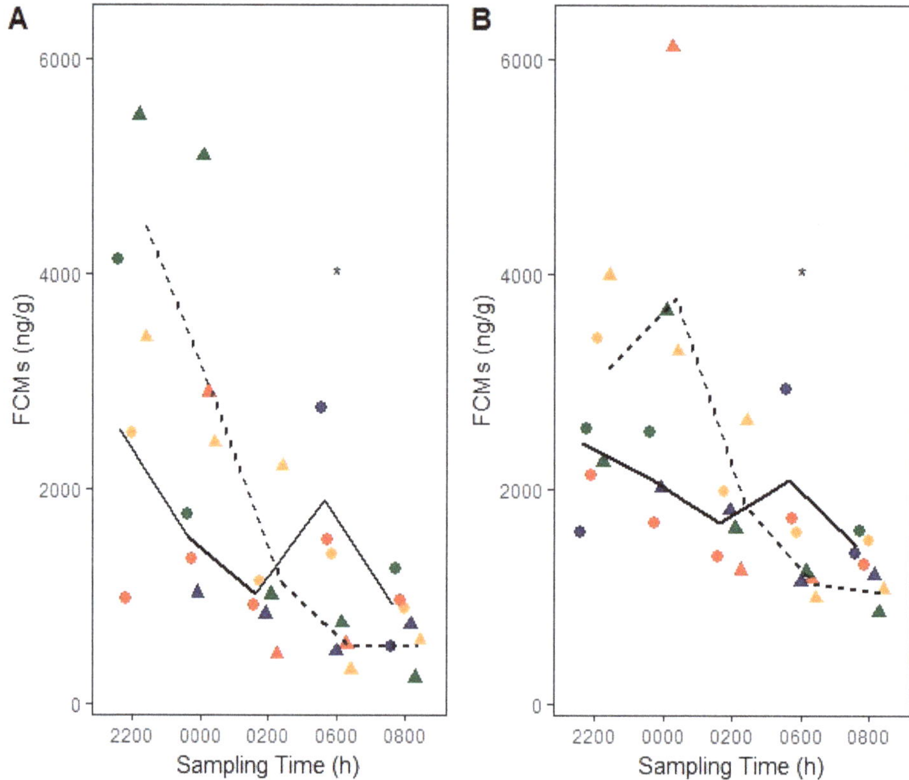

Figure 3. Corticosterone baseline fecal cortisol/corticosterone metabolites (FCMs) (solid lines and circles) versus post dexamethasone FCMs (dashed lines and triangles) in laboratory-bred deermice measured with (**A**) corticosterone enzyme immunoassay (EIA) and (**B**) group-specific EIA. Lines connect means from each sampling time. Circles and triangles indicate individual data points whereby an individual is denoted with a different color. Males are shown in blue and green colors, whereas females are shown in red and tan. Dexamethasone was administered at ~20:00 h (not shown). Asterisk (*) denotes significant differences at α = 0.05.

3.3. ACTH Stimulation Test

We found an effect of treatment for the corticosterone EIA ($F_{1, 19.22}$ = 11.94, p = 0.003; Figure 4A) where deermice had consistently higher FCM levels (7.77 ± 0.19 ln ng/g) post injection than baseline (7.27 ± 0.20 ln ng/g, post hoc, $t_{19.17}$ = 3.44, p = 0.003). We also found an effect of sampling time ($F_{4, 19.19}$ = 5.42, p = 0.004) where deermice had higher FCM levels at 22:00 h (7.84 ± 0.23 ln ng/g; $t_{19.07}$ = 4.06, p = 0.005), 00:00 h (7.74 ± 0.24 ln ng/g; $t_{19.17}$ = 3.45, p = 0.02), and 06:00 h (7.62 ± 0.22 ln ng/g; $t_{19.05}$ = 3.14, p = 0.04) compared to 08:00 h (6.99 ± 0.22 ln ng/g) regardless of treatment. Similarly, using the group-specific EIA, we found an effect of treatment ($F_{1, 20.25}$ = 14.16, p = 0.001; Figure 4B) where deermice had higher (7.99 ± 0.12 ln ng/g) FCM levels consistently post injection than baseline (7.58 ± 0.12 ln ng/g, $t_{20.16}$ = 3.75, p = 0.001; Figure 4B). Again, we also found an effect of sampling time ($F_{4, 20.28}$ = 3.42, p = 0.03) where deermice had higher FCM levels at 22:00 h (8.02 ± 0.15 ln ng/g) compared to

08:00 h (7.48 ± 0.14 ln ng/g, $t_{20.06}$ = 3.50, p = 0.02) regardless of treatment. We note that baseline FCM levels at 06:00 h represent a 2 h interval (04:00–06:00 h) whereas treatment FCM levels at the same time represent a 4 h interval (02:00–06:00 h; Table 1).

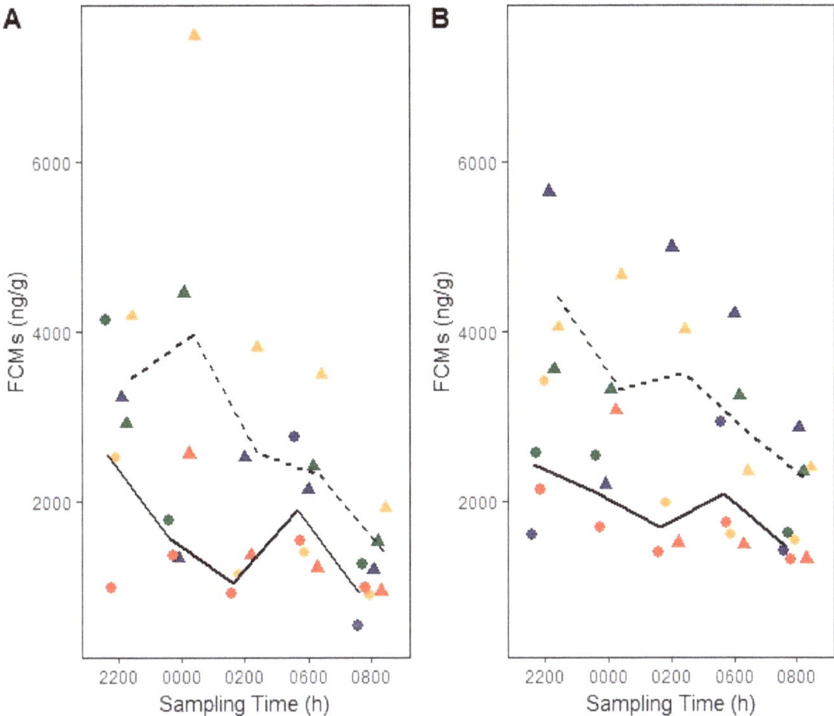

Figure 4. Corticosterone baseline fecal cortisol/corticosterone metabolites (FCMs) (solid lines and circles) versus post adrenocorticotropic hormone (ACTH) FCMs (dashed lines and triangles) in laboratory-bred deermice measured with (**A**) corticosterone enzyme immunoassay (EIA) and (**B**) group-specific EIA. Lines connect means from each sampling time. Circles and triangles indicate individual data points whereby an individual is denoted with a different color. Males are shown in blue and green colors, whereas females are shown in red and tan. ACTH was administered at ~20:00 h (not shown).

3.4. Field Validation

For field study 1 (group-specific EIA), we found that there was a marginal effect of confinement time on FCM levels ($F_{2,15}$ = 3.38, p = 0.06), which we still elected to explore because of likely biological relevance. We found no effect of sex (p = 0.56) or reproductive status (p = 0.39). Deermice confined for 4–6 h had marginally higher FCM levels (n = 6, 4 females and 2 males, 8.33 ± 0.08 ln ng/g) compared to those confined for 0–2 h (n = 7, 4 females and 3 males, 7.90 ± 0.11 ln ng/g, p = 0.07). However, deermice confined for 0–2 h had no differences in their FCM levels compared to those of deermice confined for 8–10 h (n = 7, 5 females and 2 males, p = 0.16) or 4–6 h versus 8–10 h (p = 0.85; Figure 5). For field study 2 (corticosterone EIA), we found a significant effect of confinement time on FCM levels ($F_{1,10}$ = 23.21, p = 0.001). Deermice had lower FCMs (9.12 ± 0.32 ln ng/g) when confined for 0–4 h compared to after short-term restraint and overnight confinement (n = 7, 11.10 ± 0.32 ln ng/g, t_6 = 4.82, p = 0.003; Table 2). Sex (p = 0.60) and reproductive status (p = 0.75) were not significant.

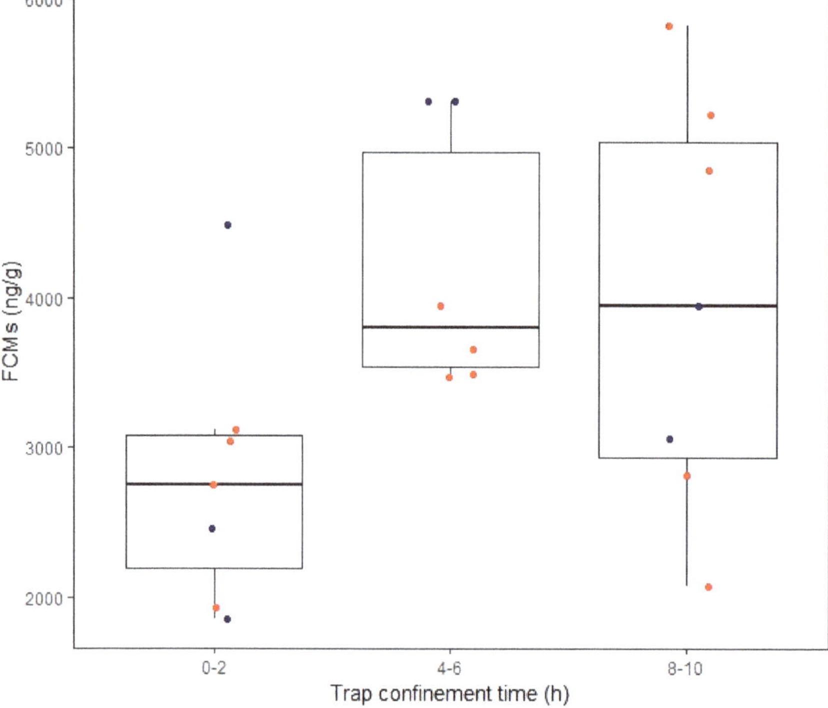

Figure 5. Fecal cortisol/corticosterone metabolites (FCMs) across different trap confinement times in free-ranging deermice. FCMs were measured with the group-specific EIA. Boxplots display the median (line), 25–75% interquartile range (boxes) and the full range (whiskers). In the 0–2 h group, there is a large outlier. Circles indicate individual data points whereby blue denotes males and red denotes females.

Table 2. Summary of demographic factors and fecal cortisol/corticosterone metabolite (FCM) values from free-ranging deermice captured in MT, USA in August 2017 for field study 2.

Sex	Reproductive [1]	FCMs at 0–4 h (ng/g)	FCMs Overnight (ng/g)	FCM Difference [2]
Male	No	21,034	49,709	28,676
Male	No	5407	76,071	70,663
Male	Yes	11,224	185,928	174,704
Male	Yes	4649	91,514	86,865
Female	Yes	37,172	37,776	604
Female	Yes	3611	41,678	38,067
Female	Yes	9704	82,299	72,595

[1] Reproductive status was determined via scrotal testes in males and presence of a perforate vagina, lactation, or pregnancy in females, [2] FCM difference was calculated by subtracting FCMs at 0–4 h from FCMs overnight.

4. Discussion

We provided evidence that validates the use of both the corticosterone and the group-specific EIAs with FCMs in deermice. In the laboratory studies, both EIAs showed a similar decrease and increase in FCMs post dexamethasone and ACTH injections, respectively. Despite the group-specific EIA's ability to detect particular corticosterone metabolites in feces, FCM values were comparable to the ones we detected with the corticosterone EIA. Field study 1 (group-specific EIA) showed that deermice had marginally higher FCM levels when confined for 4–6 h versus 0–2 h. Field study

2 (corticosterone EIA) more strongly echoed these results where deermice had higher FCMs after short-term restraint and confinement more than 0–4 h. Although we did not verify whether the stressors we used increased blood corticosterone, we do not think this affects our conclusions because of two main reasons. Firstly, the stressors we used have been known to influence blood GCs in other species (e.g., [18,35]), and secondly, many other validation studies for FCMs were successful without performing this type of verification (e.g., [15,36–38]). Although comparing between studies with different extraction and assay protocols is difficult, both EIAs we used consistently detected higher values compared to the corticosterone RIA used in previous deermouse studies (e.g., [10,11]).

4.1. Diurnal Rhythm and Sex Effects

We detected no effects of diurnal rhythm or sex on FCMs with either EIA when using only data from the third day in captivity. However, the variability in FCM data was smaller towards the end of the dark cycle compared to the beginning, suggesting that if we had FCM data from each deermouse for each time point, a significant change over time may have manifested. Regardless, we did find higher FCMs at 22:00 h compared to 08:00 h when using pooled data from the ACTH challenge for both EIAs. In fact, FCMs were also higher at 00:00 h and 06:00 h compared to 08:00 h for the corticosterone EIA. This finding from pooled data is most likely because, although treatment FCMs were relatively higher than baseline, they still showed a declining trend, similar to baseline FCMs, across the dark cycle. Previous studies with small mammals found either a presence or absence of a diurnal rhythm in FCMs (e.g., [24,25]). However, when a diurnal rhythm is found, FCM levels will typically rise before the period of highest activity and start to decrease closer to the period of inactivity, which is similar to what we found [25,39]. This reflects the dynamics of blood GCs before they appear in feces, which is governed by a species-specific time delay [4]. This delay can range from 4 h in small mammals (e.g., house mice, [15]) to ~24 h in larger mammals [20]. Similarly, [40] found no effect of sex on FCM levels in deermice, although [25] detected sex differences in house mice where females had higher FCM levels. Due to low sample size of females and males, there may have been an effect of sex on FCMs, which we were unable to detect.

4.2. Suppression of Adrenocortical Activity

We found that FCM levels decreased significantly ~10 h post dexamethasone injection with both EIAs. FCMs decreased on average by ~73% and ~43% for the corticosterone and group-specific EIAs, respectively. This however, could have happened sooner (i.e., ~8 h post injection) since we lacked FCM data at 04:00 h. Nevertheless, other rodent studies found FCM levels decreased 8-10 h post dexamethasone in house mice [25] and 10–12 h in Norway rats (*Rattus norvegicus*) [41], although injections were given during the light cycle. However, the percentage decreases we observed were lower than in Norway rats (~86%; [41]) but higher than in Columbian ground squirrels (*Urocitellus columbianus*) (~33%; [36]), both of which used the same group-specific EIA. This could suggest that a higher dexamethasone dose could be used to more strongly suppress FCMs in deermice. It is noteworthy that the group-specific EIA did detect a marginal increase in FCM levels ~4 h post dexamethasone injection (~69% average increase), most likely due to restraint/injection stress. Even if a higher dose may have resulted in a larger effect size, both EIAs tracked the expected suppression in FCMs post dexamethasone.

4.3. Stimulation of Adrenocortical Activity

FCM levels increased ~2 h post ACTH injection and remained elevated when using both EIAs. In particular, FCMs on average increased by ~65% and ~50% with the corticosterone and group-specific EIAs, respectively. Given that other rodent studies found longer time delays than 2 h post ACTH injection, such as 5–7 h in Egyptian spiny mice (*Acomys cahirinus*) [42] and 6–8 h in bank voles (*Myodes glareolus*) [43], this finding was unexpected. However, in these studies the ACTH injections were given during the light cycle in rodents that are mostly nocturnal, which could have affected time delays [15].

Regardless, [44] did find that brown lemmings (*Lemmus trimucronatus*) reached their half maxima FCM values within 2 h of capture, anesthesia, and transportation in the field. Similarly, [26] found that fecal radioactivity appeared as early as 2 h in a radiometabolism study of California mice (*Peromyscus californicus*). Nonetheless, in our study, it is still possible that capture and restraint for injection significantly decreased gut passage time [4]. The percentage increases we found are much lower than previous mammal studies. For example, [36] found ~255% increase post ACTH in Columbian ground squirrels using the group-specific EIA. However, [24] found ~56% increase in meadow voles using the group-specific EIA. Therefore, the ACTH dose we used may not have been high enough to reach a stronger effect. Regardless, the modest yet significant increase in FCM levels post ACTH injection provides validation evidence for both EIAs.

4.4. Trap-Induced Effects on FCMs

Trap confinement for 4–6 h marginally increased FCM levels in free-ranging deermice, compared to confinement for 0–2 h but not compared to 8–10 h (field study 1). Similarly, deermice confined for 0–4 h had lower FCM levels compared to additional confinement of ~7 h and after short-term restraint (field study 2). Although we cannot easily tease apart effects from restraint and trap confinement time in field study 2, the findings still provide biological validation. Because FCM levels tend to decrease shortly into the active phase [25,39], the elevations we observed after 4 h would most likely have been due to trap-induced stress. The lack of difference between 0–2 h and 8–10 h could stem from how the stressor of trap confinement remained consistent over time so that FCMs eventually returned to baseline. Alternatively, it could be that the natural decline of FCMs overnight conflicted with the increase in FCMs from trap-induced stress, thereby leading to a lower average FCMs and a larger variability in the data for the 8–10 h group (Figure 5). Similarly, [21] found no differences in FCM levels between deermice in traps for 4–8 h versus overnight, although FCM levels did continue to increase with more trap confinement in another deermouse population. Based on field study 1 findings, the lag time between corticosterone in the blood to excretion in the feces may be ~4 h during the period of highest activity (i.e., dark cycle). This is similar to what has been reported in another deermouse study that used a corticosterone RIA [21]. Although sex and reproductive status can influence stress physiology [4], we did not find any effects on FCMs from sex or reproductive status. However, this may have been due to low sample sizes, and not a limitation of the EIAs. Nevertheless, our findings suggest that trap-induced stress may affect FCM levels even within 4 h of confinement so that earlier fecal collection may better capture baseline adrenocortical activity and unmask individual heterogeneity.

4.5. Drying Effects on FCMs

Although samples oven-dried for 1 h and then lyophilized were diluted 1:10 for the corticosterone EIA, samples oven-dried for 2 h with no lyophilization had to be diluted 1:80 instead. This increase in the dilution factor could be the result of additional drying time where further alteration of FCMs can affect actual FCM levels and influence antibody binding. Similar heat effects on FCM levels were reported by [45] where autoclaving ungulate feces artificially increased FCM levels. However, an alternative reason could be the origin of the samples because those that were diluted more came from free-ranging deermice whereas those that were diluted less came from laboratory deermice on rodent chow diet. Because diet can affect FCM levels, the diet of free-ranging deermice may have led to artificially higher FCM levels [46]. However, because the group-specific EIA did not detect differences between laboratory and wild deermice (i.e., same dilution factor), it is most likely that an additional hour of oven-drying induced structural changes to FCMs that were detected by the corticosterone EIA antibody, thereby increasing FCM levels (e.g., [47]). Therefore, the drying protocol needs to remain consistent throughout a study, (e.g., for multiple samples from one individual) if valid FCM comparisons are to be made.

5. Conclusions

We analytically, physiologically, and biologically validated two EIAs for measuring FCMs in deermice. Although we used identical sample processing and extraction methods for laboratory samples, this was not the case with field samples so direct comparisons should be made with caution. Nevertheless, both field studies demonstrated similar temporal patterns, so they provided biological validation for the two EIAs. Although we acknowledge low sample sizes, our study is the first to provide validation evidence for EIAs that can quantify FCMs in deermice.

Author Contributions: Conceptualization, A.E. and R.B.; data curation, A.E.; formal analysis, A.E.; funding acquisition, A.E. and R.B.; investigation, A.E. and R.B.; methodology, A.E. and R.B.; project administration, A.E.; resources, A.E., R.P., and R.B.; software, A.E.; supervision, A.E.; validation, A.E.; visualization, A.E.; writing—original draft, A.E.; writing—review and editing, A.E., R.P., and R.B. All authors have read and agreed to the published version of the manuscript.

Funding: This research was partially funded by the Wildlife Biology program at the University of Montana, Interdisciplinary Collaborative Network, Montana Institute on Ecosystems and the Associated Students of the University of Montana.

Acknowledgments: We thank Grant McClelland and Graham Scott of McMaster University for providing us with the laboratory-bred deermice that originated from Nebraska, USA. We thank Michael Sheriff for all his advice in setting up the study, for partially funding laboratory analyses and for allowing the primary author to complete assay work at his laboratory space at Pennsylvania State University. We also thank Catharine E. Pritchard for her laboratory and editorial assistance, Ben Dantzer from the University of Michigan for shipping samples to us at Pennsylvania State University and Angela Luis from the University of Montana in Missoula for her general feedback and support during the study. Additionally, we thank the private cattle ranch owners and Montana Fish, Wildlife and Parks for allowing us access to their lands where the fieldwork took place, and all the volunteers that assisted with field and lab work at the University of Montana. Lastly, we thank the anonymous reviewers for their helpful comments on previous versions of this manuscript.

Conflicts of Interest: The authors declare no conflict of interest.

References

1. Sheriff, M.J.; Dantzer, B.; Delehanty, B.; Palme, R.; Boonstra, R. Measuring stress in wildlife: Techniques for quantifying glucocorticoids. *Oecologia* **2011**, *166*, 869–887. [CrossRef] [PubMed]
2. Dantzer, B.; Fletcher, Q.E.; Boonstra, R.; Sheriff, M.J. Measures of physiological stress: A transparent or opaque window into the status, management and conservation of species? *Conserv. Physiol.* **2014**, *2*, cou023. [CrossRef] [PubMed]
3. Touma, C.; Palme, R. Measuring fecal glucocorticoid metabolites in mammals and birds: The importance of validation. *Ann. NY Acad. Sci.* **2005**, *1046*, 54–74. [CrossRef]
4. Palme, R. Non-invasive measurement of glucocorticoids: Advances and problems. *Physiol. Behav.* **2019**, *199*, 229–243. [CrossRef] [PubMed]
5. Fanson, K.V.; Best, E.C.; Bunce, A.; Fanson, B.G.; Hogan, L.A.; Keeley, T.; Narayan, E.J.; Palme, R.; Parrott, M.L.; Sharp, T.M. One size does not fit all: Monitoring faecal glucocorticoid metabolites in marsupials. *Gen. Comp. Endocrinol.* **2017**, *244*, 146–156. [CrossRef]
6. Heistermann, M.; Palme, R.; Ganswindt, A. Comparison of different enzyme immunoassays for assessment of adrenocortical activity in primates based on fecal analysis. *Am. J. Primatol.* **2006**, *68*, 257. [CrossRef] [PubMed]
7. Schountz, T.; Prescott, J.; Cogswell, A.C.; Oko, L.; Mirowsky-Garcia, K.; Galvez, A.P.; Hjelle, B. Regulatory T cell-like responses in deer mice persistently infected with Sin Nombre virus. *Proc. Natl. Acad. Sci. USA* **2007**, *104*, 15496–15501. [CrossRef]
8. Mahalingam, S.; McClelland, G.B.; Scott, G.R. Evolved changes in the intracellular distribution and physiology of muscle mitochondria in high-altitude native deer mice. *J. Physiol. (Lond.)* **2017**, *595*, 4785–4801. [CrossRef]
9. Luis, A.D.; Douglass, R.J.; Mills, J.N.; Bjørnstad, O.N. Environmental fluctuations lead to predictability in Sin Nombre hantavirus outbreaks. *Ecology* **2015**, *96*, 1691–1701. [CrossRef]
10. Harper, J.M.; Austad, S.N. Fecal glucocorticoids: A noninvasive method of measuring adrenal activity in wild and captive rodents. *Physiol. Biochem. Zool.* **2000**, *73*, 12–22. [CrossRef]

11. Hayssen, V.; Harper, J.; DeFina, R. Fecal corticosteroids in agouti and non-agouti deer mice (*Peromyscus maniculatus*). *Comp. Biochem. Physiol. A Mol. Integr. Physiol.* **2002**, *132*, 439–446. [CrossRef]
12. Pedersen, A.B.; Greives, T.J. The interaction of parasites and resources cause crashes in a wild mouse population. *J. Anim. Ecol.* **2008**, *77*, 370–377. [CrossRef] [PubMed]
13. Fredebaugh-Siller, S.; Suski, C.; Zuckerman, Z.; Schooley, R. Ecological correlates of stress for a habitat generalist in a biofuels landscape. *Can. J. Zool.* **2013**, *91*, 853–858. [CrossRef]
14. Bradley, E.L.; Terman, C.R. A comparison of the adrenal histology, reproductive condition, and serum corticosterone concentrations of prairie deermice (*Peromyscus maniculatus bairdii*) in captivity. *J. Mammal.* **1981**, *62*, 353–361. [CrossRef]
15. Touma, C.; Sachser, N.; Möstl, E.; Palme, R. Effects of sex and time of day on metabolism and excretion of corticosterone in urine and feces of mice. *Gen. Comp. Endocrinol.* **2003**, *130*, 267–278. [CrossRef]
16. Wielebnowski, N.; Watters, J. Applying fecal endocrine monitoring to conservation and behavior studies of wild mammals: Important considerations and preliminary tests. *Isr. J. Ecol. Evol.* **2007**, *53*, 439–460. [CrossRef]
17. Delehanty, B.; Boonstra, R. Impact of live trapping on stress profiles of Richardson's ground squirrel (*Spermophilus richardsonii*). *Gen. Comp. Endocrinol.* **2009**, *160*, 176–182. [CrossRef]
18. Bosson, C.; Islam, Z.; Boonstra, R. The impact of live trapping and trap model on the stress profiles of North American red squirrels. *J. Zool.* **2012**, *288*, 159–169. [CrossRef]
19. Millspaugh, J.J.; Washburn, B.E. Use of fecal glucocorticoid metabolite measures in conservation biology research: Considerations for application and interpretation. *Gen. Comp. Endocrinol.* **2004**, *138*, 189–199. [CrossRef]
20. Palme, R.; Rettenbacher, S.; Touma, C.; El-Bahr, S.; Möstl, E. Stress hormones in mammals and birds: Comparative aspects regarding metabolism, excretion, and noninvasive measurement in fecal samples. *Ann. NY Acad. Sci.* **2005**, *1040*, 162–171. [CrossRef]
21. Harper, J.M.; Austad, S.N. Effect of capture and season on fecal glucocorticoid levels in deer mice (*Peromyscus maniculatus*) and red-backed voles (*Clethrionomys gapperi*). *Gen. Comp. Endocrinol.* **2001**, *123*, 337–344. [CrossRef] [PubMed]
22. Montiglio, P.; Pelletier, F.; Palme, R.; Garant, D.; Réale, D.; Boonstra, R. Noninvasive monitoring of fecal cortisol metabolites in the eastern chipmunk (*Tamias striatus*): Validation and comparison of two enzyme immunoassays. *Physiol. Biochem. Zool.* **2012**, *85*, 183–193. [CrossRef] [PubMed]
23. Fletcher, Q.E.; Boonstra, R. Impact of live trapping on the stress response of the meadow vole (*Microtus pennsylvanicus*). *J. Zool.* **2006**, *270*, 473–478. [CrossRef]
24. Edwards, P.D.; Dean, E.K.; Palme, R.; Boonstra, R. Assessing space use in meadow voles: The relationship to reproduction and the stress axis. *J. Mammal.* **2019**, *100*, 4–12. [CrossRef]
25. Touma, C.; Palme, R.; Sachser, N. Analyzing corticosterone metabolites in fecal samples of mice: A noninvasive technique to monitor stress hormones. *Horm. Behav.* **2004**, *45*, 10–22. [CrossRef]
26. Harris, B.N.; Saltzman, W.; de Jong, T.R.; Milnes, M.R. Hypothalamic–pituitary–adrenal (HPA) axis function in the California mouse (*Peromyscus californicus*): Changes in baseline activity, reactivity, and fecal excretion of glucocorticoids across the diurnal cycle. *Gen. Comp. Endocrinol.* **2012**, *179*, 436–450. [CrossRef]
27. Hare, J.F.; Ryan, C.P.; Enright, C.; Gardiner, L.E.; Skyner, L.J.; Berkvens, C.N.; Anderson, W.G. Validation of a radioimmunoassay-based fecal corticosteroid assay for Richardson's ground squirrels *Urocitellus richardsonii* and behavioural correlates of stress. *Curr. Zool.* **2014**, *60*, 591–601. [CrossRef]
28. Fairbairn, D.J. The spring decline in deer mice: Death or dispersal? *Can. J. Zool.* **1977**, *55*, 84–92. [CrossRef]
29. Mills, J.N.; Yates, T.L.; Childs, J.E.; Parmenter, R.R.; Ksiazek, T.G.; Rollin, P.E.; Peters, C. Guidelines for working with rodents potentially infected with hantavirus. *J. Mammal.* **1995**, *76*, 716–722. [CrossRef]
30. R Core Team. *R: A Language and Environment for Statistical Computing*; R Foundation for Statistical Computing: Vienna, Austria, 2018.
31. RStudio Team. *RStudio: Integrated Development for R*; RStudio, Inc.: Boston, MA, USA, 2015.
32. Bates, D.; Maechler, M.; Bolker, B.; Walker, B. Fitting Linear Mixed-Effects Models using lme4. *J. Stat. Soft.* **2015**, *67*, 1–48. [CrossRef]
33. Kuznetsova, A.; Brockhoff, P.B.; Christensen, R.H.B. lmerTest Package: Tests in Linear Mixed Effects Models. *J. Stat. Soft.* **2017**, *82*, 1–26. [CrossRef]

34. Lenth, R. Emmeans: Estimated Marginal Means aka Least-Squares Means. 2018. R Package Version 1.3.1. Available online: https://CRAN.R-project.org/package=emmeans (accessed on 1 May 2020).
35. Siswanto, H.; Hau, J.; Carlsson, H.E.; Goldkuhl, R.; Abelson, K.S. Corticosterone concentrations in blood and excretion in faeces after ACTH administration in male Sprague-Dawley rats. *In Vivo* **2008**, *22*, 435–440. [PubMed]
36. Bosson, C.O.; Palme, R.; Boonstra, R. Assessment of the stress response in Columbian ground squirrels: Laboratory and field validation of an enzyme immunoassay for fecal cortisol metabolites. *Physiol. Biochem. Zool.* **2009**, *82*, 291–301. [CrossRef]
37. Sheriff, M.J.; Bosson, C.O.; Krebs, C.J.; Boonstra, R. A non-invasive technique for analyzing fecal cortisol metabolites in snowshoe hares (*Lepus americanus*). *J. Comp. Physiol. B* **2009**, *179*, 305–313. [CrossRef] [PubMed]
38. Dantzer, B.; McAdam, A.G.; Palme, R.; Fletcher, Q.E.; Boutin, S.; Humphries, M.M.; Boonstra, R. Fecal cortisol metabolite levels in free-ranging North American red squirrels: Assay validation and the effects of reproductive condition. *Gen. Comp. Endocrinol.* **2010**, *167*, 279–286. [CrossRef] [PubMed]
39. Bauer, B.; Palme, R.; Machatschke, I.H.; Dittami, J.; Huber, S. Non-invasive measurement of adrenocortical and gonadal activity in male and female guinea pigs (*Cavia aperea f. porcellus*). *Gen. Comp. Endocrinol.* **2008**, *156*, 482–489. [CrossRef] [PubMed]
40. Harper, J.M.; Austad, S.N. Fecal corticosteroid levels in free-living populations of deer mice (*Peromyscus maniculatus*) and southern red-backed voles (*Clethrionomys gapperi*). *Am. Midl. Nat.* **2004**, *152*, 400–409. [CrossRef]
41. Lepschy, M.; Touma, C.; Hruby, R.; Palme, R. Non-invasive measurement of adrenocortical activity in male and female rats. *Lab. Anim.* **2007**, *41*, 372–387. [CrossRef]
42. Nováková, M.; Palme, R.; Kutalová, H.; Janský, L.; Frynta, D. The effects of sex, age and commensal way of life on levels of fecal glucocorticoid metabolites in spiny mice (*Acomys cahirinus*). *Physiol. Behav.* **2008**, *95*, 187–193. [CrossRef]
43. Sipari, S.; Ylönen, H.; Palme, R. Excretion and measurement of corticosterone and testosterone metabolites in bank voles (*Myodes glareolus*). *Gen. Comp. Endocrinol.* **2017**, *243*, 39–50. [CrossRef]
44. Fauteux, D.; Gauthier, G.; Berteaux, D.; Bosson, C.; Palme, R.; Boonstra, R. Assessing stress in Arctic lemmings: Fecal metabolite levels reflect plasma free corticosterone levels. *Physiol. Biochem. Zool.* **2017**, *90*, 370–382. [CrossRef] [PubMed]
45. Millspaugh, J.J.; Washburn, B.E.; Milanick, M.A.; Slotow, R.; van Dyk, G. Effects of heat and chemical treatments on fecal glucocorticoid measurements: Implications for sample transport. *Wildl. Soc. Bull.* **2003**, *31*, 399–406.
46. Goymann, W. On the use of non-invasive hormone research in uncontrolled, natural environments: The problem with sex, diet, metabolic rate and the individual. *Methods Ecol. Evol.* **2012**, *3*, 757–765. [CrossRef]
47. Lexen, E.; El-Bahr, S.; Sommerfeld-Stur, I.; Palme, R.; Möstl, E. Monitoring the adrenocortical response to disturbances in sheep by measuring glucocorticoid metabolites in the faeces. *Wien. Tierarztl. Monat.* **2008**, *95*, 64

Article

Blood L-Lactate Concentration as an Indicator of Outcome in Roe Deer (*Capreolus capreolus*) Admitted to a Wildlife Rescue Center

Elena Di Lorenzo [1], Riccardo Rossi [2], Fabiana Ferrari [2], Valeria Martini [1] and Stefano Comazzi [1,*]

[1] Department of Veterinary Medicine, University of Milan, 26900 Lodi, Italy; elena.dilorenzo@studenti.unimi.it (E.D.L.); valeria.martini@unimi.it (V.M.)

[2] Piacenza Wildlife Rescue Center, 29120 Niviano di Rivergano, Italy; info@piacenzawildlife.org (R.R.); fabi1981@virgilio.it (F.F.)

* Correspondence: stefano.comazzi@unimi.it; Tel.: +39-02-503-34529

Received: 27 May 2020; Accepted: 17 June 2020; Published: 20 June 2020

Simple Summary: Roe deer are among the most frequent wild animals admitted to rescue centers in Italy. Reasons for admission include trauma, predation, starvation, and imprinting, with the final aim of hospitalization being full recovery and release into the wild for all cases. An accurate triage procedure is vital for predicting the outcome, to avoid unnecessary time spent in captivity, and an excessive allocation of time and resources on animals with a minimal chance of recovery. Since lactacidosis is often associated with death in hospitalized animals, and has been associated with poor prognosis in humans and domestic animals, we proposed an evaluation of blood L-lactate using a rapid whole blood test in order to predict the outcome of hospitalized roe deer. A cut-off of 10.2 mmol/L was identified as the best, in order distinguish animals with minimal chance of surviving and release. For these animals, humane euthanasia should be considered as an option.

Abstract: Roe deer (*Capreolus capreolus*) are among the most frequent patients of rescue centers in Italy. Three outcomes are possible: natural death, euthanasia, or treatment and release. The aim of the present study is to propose blood L-lactate concentration as a possible prognostic biomarker that may assist veterinarians in the decision-making process. Sixty-three roe deer, admitted to one rescue center in the period between July 2018 and July 2019, were sampled and divided into 4 groups according to their outcome: (1) spontaneous death (17 cases), (2) humanely euthanized (13 cases), (3) fully recovered and released (13 cases), and (4) euthanized being unsuitable for release (20 cases). In addition, blood samples from 14 hunted roe deer were analyzed as controls. Whole blood lactate concentrations were measured with a point of care lactate meter. Differences among groups were close to statistical significance ($p = 0.51$). A cut-off value of 10.2 mmol/L was identified: all the animals with higher values died or were humanely euthanized. The results suggest that roe deer with lactatemia higher than 10.2 mmol/L at admission, have a reduced prognosis for survival during the rehabilitation period, regardless of the reason for hospitalization and the injuries reported. Therefore, humane euthanasia should be considered for these animals.

Keywords: roe deer; blood parameters; prognostic factors; blood lactate concentration; biomarkers

---

1. Introduction

The roe deer (*Capreolus capreolus*) is an artiodactyl mammal belonging to the Odocoileinae subfamily—the only Euro-Asian member—and is the smallest European cervid. Roe deer are widely distributed in Europe, with the exception of northern Scandinavia and some of the islands, notably Iceland, Ireland, and the Mediterranean Sea islands. In Italy, an estimated population of more than 426,000 roe deer has been reported [1]. Roes are among the most common cervid species in Italy [2], thus they are often recovered by wildlife rescue centers. The reasons for their hospitalization are quite variable and principally include traumas (car accident or lawnmower trauma), predation, starvation, and imprinted animals, classed as being dangerous to humans or fenced into urban areas. There are three possible outcomes for hospitalized animals in a wildlife rescue center: (1) natural death, (2) euthanasia, and (3) treatment, rehabilitation, and release into the wild. An accurate and rapid triage is crucial for the veterinary staff of the centers in order to avoid prolonged treatment of animals with minimal chances of release.

Biomarkers are specific tests used to monitor normal or disease processes [3]. Such laboratory tests may help veterinary clinicians with challenging decisions regarding initial triage and management in an objective fashion. One medical complication in rescued wild animals is metabolic acidosis due to lactacidemia related to trauma, shock, capture stress, and myopathy [4]. Thus, we hypothesize that the concentration of L-lactate in blood, from roe deer admitted to a wildlife rescue center, may be used to predict outcome, irrespective of the cause of injury. The aim of the present study was to test our hypothesis and select a suitable cutoff value for L-lactate concentration in the blood of injured roe deer, to identify animals that will not survive rehabilitation and release.

2. Materials and Methods

For the purposes of the present research, blood samples from roe deer of different sexes and ages referred to the Piacenza Wildlife Rescue Center, from July 2018 to July 2019, were collected. The reasons for admission varied.

All animals were sampled for diagnostic purposes in order to better frame the clinical case (except for blood samples from the control group that were taken from hunted dead animals immediately after shooting). According to the guidelines of the authors' institution, formal approval from the Ethical Committee was not required (EC decision 29 October 2012, renewed with the protocol n° 02–2016).

The individuals included in the study are divided into classes depending on sex and age, which were determined based on morphological aspects. As far as age is concerned, three different classes can be identified: fawns, yearlings, and adults. Yearlings can be recognized by their light build and slender body. Young males have antlers which may have up to two tines, whereas females have a slender profile, as their abdomen is not relaxed by gestation. A more precise evaluation of age was obtained following a dental examination to determine the state of eruption and wearing down of the teeth.

All admitted roe deer were divided into 4 groups according to their final outcome: (Group 1) spontaneous death during the recovery period, (Group 2) euthanized for welfare reasons (i.e., animals near death where euthanasia was considered as the only possible option to reduce pain and suffering and), (Group 3) fully recovered and released roe deer, and (Group 4) animals euthanized for other reasons (i.e., animals that were judged unable to survive rehabilitation and release back into the wild). In addition, as a control (Group 5), blood samples from 14 hunted roe deer, from the same area, were taken from the cardiac cavity immediately after shooting death.

In general, a manual restraint during the recovery period (animals blindfolded and hogtied) was preferred to sedation, as chemical restraint would invalidate appropriate triage of polytraumatized animals. In the rare cases when sedation was performed, two different protocols were used: dexmedetomidine (10 µg/kg, Dexdomitor, Zoetis, Rome, Italy) + ketamine (2 mg/kg, Ketavet100, MSD) + methadone (0.2 mg/kg, Semfortan, Dechra, Turin, Italy), if the roe deer require surgery, or dexmedetomidine (10 µg/kg) + ketamine (2 mg/kg) + butorphanol (0.2 mg/kg, Dolorex, Intervet, Peschiera Borromeo, MI, Italy), if only chemical restraints, diagnostic procedures and minor surgery were required.

All samples were taken from the cephalic vein within 2 h of arrival to the Center, put into EDTA coated tubes, and L-lactate concentrations were evaluated within 10 min from sampling by means of a portable point of care lactate meter (Accutrend Plus, Roche, Monza, Italy). The instrument measures the concentration of L-lactate in whole blood using a reactive strip via a bench top clinical chemistry analyzer system, which has been validated in cattle as reliable and linear up to 16.6 mmol/L [5]. This method was preferred to classical plasma spectrophotometric evaluation using a clinical chemistry analyzer since it is portable, practical, rapid, and cheap and it does not require plasma separation, which is sometimes difficult in the absence of an equipped laboratory.

Statistic Analysis

Descriptive statistics were calculated. Data from hunted deer were used to calculate reference ranges according to the official guidelines [6]. The Dixon method was applied to identify and eliminate the outliers [7].

A Shapiro–Wilk test was performed to assess normal distribution of data among outcome groups. A Kruskal–Wallis test was performed to assess possible differences in L-lactate concentration among outcome groups. Contingency tables were prepared, and the Pearson's chi-squared test was performed to assess possible differences in sex, age, and cause of admission among outcome groups.

ROC curves were drawn, and coordinates were used to determine the cutoff value of L-lactate concentration, having 100% specificity and the highest sensitivity in the identification of animals with a negative outcome. These criteria were applied to select the cutoff value, since we aimed to identify animals with a negative outcome and avoid false positive results.

At first, survivors and animals who encountered spontaneous death were included in the ROC curve analysis. Thereafter, humanely euthanized animals (group 2) were also included and grouped together with those who died spontaneously.

All statistical analyses were performed by means of a dedicated software (SPSS 20.0 for Windows). Significance was set at $p \leq 0.05$ for all analyses.

3. Results

A total of 77 blood samples with L-lactate concentrations were available and divided into five different groups: (1) spontaneous death (17 cases), (2) humanely euthanized (13 cases), (3) fully recovered and released (13 cases), (4) euthanized for other reasons (20 cases), and (5) control animals (14 cases). A total of 30 females and 47 males, of which 38 adults, 16 subadults (10–23 months old) and 23 fawns (less than 10 months old), were tested. Among hospitalized roe deer, the prevalent causes of admission were trauma (45 cases), followed by predation (9 cases), imprinted animals (1 case) and other causes (8 cases). The four groups were homogeneous as for gender, age, and cause of admission ($p > 0.05$ for all analyses).

Results of blood lactate concentrations in different groups of roe deer are reported in Table 1 and Figure 1. Differences among groups were not significant ($p = 0.051$).

Table 1. Results of descriptive statistics on blood L-lactate concentrations among different groups of roe deer. 1 No statistical differences were detected amongst groups ($p = 0.51$). # number of cases.

Group	Whole Blood L-lactate Concentration (mmol/L)			Cases with L-lactate > 10.2 mmol/L
	Median	Min	Max	# (%)
1: spontaneous death ($n = 17$)	9.5	2.1	25.0	8/17 (47.1%)
2: euthanized for welfare reasons ($n = 13$)	9.0	1.1	24.3	5/13 (38.5%)
3: recovered and released ($n = 13$)	7.1	1.7	10.0	0/13 (0%)
4: euthanized for other reasons ($n = 20$)	4.3	0.8	15.8	5/20 (25%)
5: hunted deer ($n = 14$)	4.3	2.7	15.3	1/14 (7.1%)

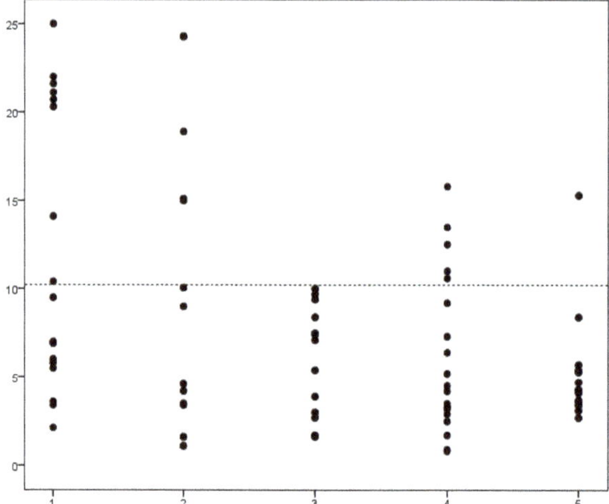

Figure 1. Dotplot showing results of blood L-lactate concentration in 77 roe deer. *X*-axis, outcome after hospitalization: (Group 1) spontaneously dead; (Group 2) euthanized for welfare reasons; (Group 3) recovered and released; (Group 4) euthanized for other reasons; (Group 5) hunted deer. *Y*-axis, L-lactate concentration (mmol/L). Dotted line: cutoff selected based on ROC curve coordinates, having a 100% specificity in discriminating between Group 1 and 3.

The reference range obtained after exclusion of outliers in our control group was 2.7–5.7 mmol/L.

Based on ROC curve coordinates, an L-lactate cutoff of 10.2 mmol/L was selected as the best to classify survivors and animals having encountered spontaneous death, with a 100% specificity and 47.1% sensitivity in detecting animals with a negative outcome.

When animals euthanized for welfare reasons (Group 2) were included in the analysis, an L-lactate cutoff of 10.0 mmol/L was selected as the best one, with a 100% specificity and a 46.7% sensitivity in detecting animals with a negative outcome. The aforementioned cutoff of 10.2 had a 100% specificity and a 43.3% sensitivity.

4. Discussion

Blood L-lactate derives from anaerobic glycolysis and in physiologic conditions, a production of 0.8 mmol/h/kg has been demonstrated, in humans, to contribute to obtain a normal plasma concentration of less than 1 mmol/L [8]. Concentration may increase with increased anaerobic metabolism due to many different causes, including tissue hypoxia, shock, increased gluconeogenesis, sepsis, anemia, or muscular overwork and damage. Blood lactate concentration has been proposed as a possible prognostic test in emergency conditions in both humans and animals [3,9]. A reduced prognosis in critical human patients has been reported with persistent high lactacidemia [9]. Furthermore, another study [10] suggested that lactacidemia higher than 4.5 mmol/L is correlated to death in 78% of critical patients with emergency conditions. Similarly, blood L-lactate concentration influences prognosis in horses with complicated colic [11], and in critically ill neonatal foals [12]. The use of a portable point of care lactate meter has also been validated in equine medicine [13], as a useful support to veterinary clinical decision making.

In wild cervids, plasma L-lactate has been evaluated as part of a laboratory panel to check the effects of capture using different methods [14,15], in different conditions [16], using different immobilizing drugs [17,18] or premedication [19]; however, to the best of our knowledge, no data about the possible role of plasma lactate concentration, to define the health status and predict the outcome of captured animals, have been published. Reference intervals for free ranging wild animals are often lacking.

In the present study, the analysis of samples from 14 hunted roe deer (control Group 5) provided some reference to compare the concentration of plasma L-lactate from recovered animals. All roes in the study derived from the same area of study and were killed by selector hunters with a single shot when they were calm and minimally alert; blood samples were immediately taken via intracardiac puncture after being shot and were rapidly transported to the center for analysis. For these reasons, we consider this group as an adequate control for defining plasma L-lactate concentration in free ranging roe deer in the study area. However, the authors acknowledge that these control animals' elevated L-lactate levels were likely associated with acute death, and that these values do not represent those of normal, healthy roe deer. In addition, cardiac blood could have a lower L-lactate concentration in comparison with venous peripheral blood, especially if oxygenated. In comparison, the reference range obtained in our control group was 2.7–5.7 mmol/L, lower than previously reported for 22 trapped roe deer (5.1–18.9 mmol/L) [11]. This supports the hypothesis that capture itself induces an increase in plasma L-lactate concentration. Therefore, for all the above-mentioned reasons, in the present research, we preferred comparing values from rescued roe deer with different outcomes, instead of using this reported reference interval. Interestingly, in this control group, a couple of samples showed higher L-lactate concentrations than the reference interval (8.4 and 15.3 mmol/L), and in one case, it was also higher than the prognostic cutoff values identified. This result supports the high variability of this parameter and confirms that plasma L-lactate concentration cannot be used as a stand-alone parameter to predict prognosis, especially when interpreting single admission L-lactate values, rather it should be considered in the context of clinical examination. In addition, the higher values in some of the control deer, could suggest that the control group was not 100% fit and healthy and/or had suffered from stress, dehydration, blood loss or other causes of elevated L-lactate, since we do not have any information about timeframes between the shooting and actual death of each animal.

If we considered hospitalized roe deer, in which the modality of rescue, immobilization and handling are consistent, we found that the overall ranges of plasma L-lactate concentration are similar to those reported in previous studies for trapped animals [15,19], further supporting the possible effect of capture stress.

The cutoff value of 10.2 mmol/L was chosen as a negative prognostic indicator, representing deer that are most likely destined to spontaneous death or require humane euthanasia. We selected a cutoff value with 100% specificity, where few, if any, animals with L-lactate above this level would survive, to avoid under-treatment or euthanasia of animals with a possible chance of surviving. This high cutoff value is responsible for the low sensitivity found. The limited sensitivity allowed us to identify less than half of the deer with a poor prognosis, but we cannot state that the low L-lactate concentrations are necessarily associated with a good outcome. This choice was made since, from a clinical point of view, we preferred to avoid any under-treatment (or euthanasia) in roe deer with a chance of survival. We are aware that this value may induce an overestimation of the chances of survival for some animals, but we consider it an acceptable risk from an ethical perspective. If we also apply this cut-off value to Group 4 (animals euthanized for other reasons) we found that 5 cases out of 20 (25%) showed L-lactate concentrations higher than the value of 10.2 mmol/L, thus supporting euthanasia in these animals. The median concentration of L-lactate found in non-surviving deer compares well (although it is a little lower) with those found at admission in one study of horses with colonic volvulus (non-survivor: 9.1 mmol/L) [20] and in another of dogs with gastric dilation and volvulus (non-survivor: 7.9 mmol/L) [21]. However, the cutoff value selected in the present study is higher than those used in the above-mentioned studies (6.0 mmol/L and 7.4 mmol/L, respectively) since, as already mentioned, we made the choice of using 100% specificity for predicting a bad outcome in roe deer. To our knowledge, no specific studies on the prognostic value of L-lactate in roe deer or other cervids are available to date.

The present study has some limitations. First, the limited number of cases did not allow us to stratify cases according to age and cause of admission. Second, plasma L-lactate concentration was evaluated at arrival at the rescue center, even though the time from being found to arrival at the center was variable according to different aspects, which may have biased our results. However, blood withdrawal was not always possible at the time of animal discovery or injury, (since it would require veterinary professional intervention in all the cases) and therefore, we chose to use the time of first admission to the center to best standardize and compare results. Third, few of the roe deer were sedated to allow for easier transportation and to decrease stress. However, this procedure was not standardized, since it depended upon the clinical condition, type of problem, and the presence of a veterinarian, and it was not possible to revisit which cases received sedation. Premedication with acepromazine has been reported to possibly decrease capture stress and myopathy, and reduce L-lactate concentration, thus increasing survival [19]. However, to the best of our knowledge, no data are available regarding the possible effects sedation could have on L-lactate concentration. Importantly, in terms of prognosis, this would not affect the value of L-lactate concentration for predicting outcome.

In addition, in the present work we also added a control group from hunted roe deer who were shot dead as a comparison. The authors are aware that this group is probably not ideal for the creation of a reference interval, since it is biased by many different conditions and also influenced by different methods of blood withdrawal (cardiac in control deer vs peripheral blood in other deer). However, obtaining a reference interval in healthy wild animals is often challenging, since many blood analyte concentrations are strongly influenced by capture stress, and the use of a control group kept in captivity is probably not adequate as a comparison. Despite these limitations, the authors speculate that results from hunted deer may be useful for comparison since they were lower than the trapped roe deer reference value.

Lastly, our work only included L-lactate values at admission and did not evaluate serial L-lactate in these animals, which has been reported to be more valuable than a single admission value when determining a prognosis [22,23].

Future studies should assess possible fluctuations of L-lactate concentrations over the time between admission and release and their possible prognostic role. In addition, it would be interesting to validate the same test and define accurate cutoff values for other types of samples, such as capillary blood taken from an ear-edge prick, which is minimally invasive and does not require a veterinary professional. Considering the limited volume of blood required for the test (50 μL), such a small amount of blood could be sufficient for such testing.

5. Conclusions

The data obtained in the present research support the use of plasma lactate concentration at admission to predict outcome of recovered roe deer, independently from the cause of recovery. Plasma lactate may be easily evaluated using a point of care portable analyzer in a very rapid, cheap, and feasible way. The results obtained may be interpreted by veterinary professionals in conjunction with other clinical information to predict prognosis, thus avoiding directing excessive efforts and resources on untreatable animals, and reducing over- and mis-treatments. Furthermore, our results will help guide euthanasia as a humane solution for wild animals facing little chance of release back into the wild.

Author Contributions: Conceptualization R.R., S.C.; Investigation E.D.L., R.R., F.F.; Statistical analysis V.M.; Writing, original draft preparation S.C., E.D.L.; Writing, review and editing, R.R., F.F., V.M. All authors have read and agreed to the published version of the manuscript.

Funding: This research received no external funding.

Acknowledgments: The authors wish to thank Charlotte Rendina for revising English language.

Conflicts of Interest: The authors declare no conflict of interest.

References

1. Focardi, S.; Montanaro, P.; La Morgia, V.; Riga, F. *Piano D'azione Nazionale per il Capriolo Italico; (Capreolus capreolus Italicus)*; Quaderni di conservazione della natura; Ministero dell'Ambiente e ISPRA publ.: Rome, Italy, 2009; Volume 31.
2. Riga, F.; Genghini, M.; Cascione, C.; Di Luzio, P. *Impatto Degli Ungulati Sulle Culture Agricole a Forestali Proposta per Linee Guida Nazionali*; Manuali e Linee Guida ISPRA, ISPRA publ.: Rome, Italy, 2011; Volume 68.
3. Radcliffe, R.M.; Buchanan, B.R.; Cook, V.L.; Divers, T.J. The clinical value of whole blood point-of-care biomarkers in large animal emergency and critical care medicine. *J. Vet. Emerg. Crit. Care* **2015**, *25*, 138–151. [CrossRef] [PubMed]
4. Mutlow, A. BSAVA Manual of Wildlife Casualties. *J. Zoo Wildl. Med.* **2004**, *35*, 255. [CrossRef]
5. Karapinar, T.; Kaynar, O.; Hayirli, A.; Kom, M. Evaluation of 4 point-of-care units for the determination of blood l-lactate concentration in cattle. *J. Vet. Intern. Med.* **2013**, *27*, 1596–15604. [CrossRef] [PubMed]
6. Friedrichs, K.R.; Harr, K.E.; Freeman, K.P.; Szladovits, B.; Walton, R.M.; Barnhart, K.F.; Blanco-Chavez, J. American Society for Veterinary Clinical Pathology ASVCP reference interval guidelines: Determination of de novo reference intervals in veterinary species and other related topics. *Vet. Clin. Pathol.* **2012**, *41*, 441–453. [CrossRef]
7. Dixon, W.J. Processing Data for Outliers. *Biometrics* **1953**, *9*, 74. [CrossRef]
8. Weil, M.H.; Afifi, A.A. Experimental and Clinical Studies on Lactate and Pyruvate as Indicators of the Severity of Acute Circulatory Failure (Shock). *Circulation* **1970**, *41*, 989–1001. [CrossRef]
9. Cheung, P.Y.; Finer, N.N. Plasma lactate concentration as a predictor of death in neonates with severe hypoxemia requiring extracorporeal membrane oxygenation. *J. Pediatr.* **1994**, *125*, 763–768. [CrossRef]
10. Hatherill, M.; McIntyre, A.G.; Wattie, M.; Murdoch, I.A. Early hyperlactataemia in critically ill children. *Intensiv. Care Med.* **2000**, *26*, 314–318. [CrossRef]
11. Yamout, S.Z.; Nieto, J.E.; Beldomenico, P.M.; DeChant, J.E.; Lejeune, S.; Snyder, J.R. Peritoneal and Plasma d-lactate Concentrations in Horses with Colic. *Vet. Surg.* **2011**, *40*. [CrossRef]
12. Borchers, A.; Wilkins, P.A.; Marsh, P.M.; Axon, J.E.; Read, J.; Castagnetti, C.; Pantaleon, L.; Clark, C.; Qura'N, L.; Belgrave, R.; et al. Association of admission L-lactate concentration in hospitalised equine neonates with presenting complaint, periparturient events, clinical diagnosis and outcome: A prospective multicentre study. *Equine Vet. J.* **2012**, *44*, 57–63. [CrossRef]
13. Nieto, J.E.; DeChant, J.E.; Le Jeune, S.S.; Snyder, J.R. Evaluation of 3 Handheld Portable Analyzers for Measurement ofL-Lactate Concentrations in Blood and Peritoneal Fluid of Horses with Colic. *Vet. Surg.* **2014**, *44*, 366–372. [CrossRef] [PubMed]
14. Huber, N.; Vetter, S.G.; Evans, A.L.; Kjellander, P.; Küker, S.; Bergvall, U.A.; Arnemo, J.M. Quantifying capture stress in free ranging European roe deer (Capreolus capreolus). *BMC Vet. Res.* **2017**, *13*, 127.
15. Boesch, J.M.; Boulanger, J.R.; Curtis, P.D.; Erb, H.N.; Ludders, J.W.; Kraus, M.S.; Gleed, R.D. Biochemical Variables in Free-Ranging White-Tailed Deer (Odocoileus Virginianus) After Chemical Immobilization in Clover Traps or via Ground-Darting. *J. Zoo Wildl. Med.* **2011**, *42*, 18–28. [CrossRef] [PubMed]
16. Costa, G.L.; Nastasi, B.; Musicò, M.; Spadola, F.; Morici, M.; Cucinotta, G.; Interlandi, C. Influence of Ambient Temperature and Confinement on the Chemical Immobilization of Fallow Deer (Dama dama). *J. Wildl. Dis.* **2017**, *53*, 364–367. [CrossRef] [PubMed]
17. Haga, H.A.; Wenger, S.; Hvarnes, S.; Os, O.; Rolandsen, C.M.; Solberg, E.J. Plasma lactate concentrations in free-ranging moose (Alces alces) immobilized with etorphine. *Vet. Anaesth. Analg.* **2009**, *36*, 555–561. [CrossRef] [PubMed]
18. Evans, A.L.; Fahlman, A.; Ericsson, G.; Haga, H.A.; Arnemo, J. Physiological Evaluation of Free-Ranging Moose (Alces Alces) Immobilized with Etorphine-Xylazine-Acepromazine in Northern Sweden. *Acta Vet. Scand.* **2012**, *54*, 77. [CrossRef] [PubMed]
19. Montane, J.; Marco, I.; López-Olvera, J.R.; Rossi, L.; Manteca, X.; Lavín, S. Effect of acepromazine on the signs of capture stress in captive and free-ranging roe deer (Capreolus capreolus). *Vet. Rec.* **2007**, *160*, 730–738. [CrossRef] [PubMed]
20. Johnston, K.; Holcombe, S.J.; Hauptman, J.G. Plasma lactate as a predictor of colonic viability and survival after 360 degrees volvulus of the ascending colon in horses. *Vet. Surg.* **2007**, *36*, 563–567.

21. Beer, K.A.S.; Syring, R.S.; Drobatz, K.J. Evaluation of plasma lactate concentration and base excess at the time of hospital admission as predictors of gastric necrosis and outcome and correlation between those variables in dogs with gastric dilatation-volvulus: 78 cases (2004–2009). *J. Am. Vet. Med. Assoc.* **2013**, *242*, 54–58. [CrossRef]
22. Borchers, A.; Wilkins, P.A.; Marsh, P.M.; Axon, J.E.; Read, J.; Castagnetti, C.; Pantaleon, L.; Clark, C.; Qura'N, L.; Belgrave, R.; et al. Sequential L-lactate concentration in hospitalised equine neonates: A prospective multicentre study. *Equine Vet. J.* **2013**, *45*, 2–7. [CrossRef]
23. Tennent-Brown, B.; Wilkins, P.; Lindborg, S.; Russell, G.; Boston, R. Sequential Plasma Lactate Concentrations as Prognostic Indicators in Adult Equine Emergencies. *J. Vet. Intern. Med.* **2010**, *24*, 198–205. [CrossRef] [PubMed]

Article

Comparison of the Glucocorticoid Concentrations between Three Species of *Lemuridae* Kept in a Temporary Housing Facility

Martina Volfova *, Zuzana Machovcova, Eva Voslarova, Iveta Bedanova and Vladimir Vecerek

Department of Animal Protection and Welfare and Veterinary Public Health, Faculty of Veterinary Hygiene and Ecology, University of Veterinary and Pharmaceutical Sciences Brno, 612 42 Brno, Czech Republic; machovcovaz@vfu.cz (Z.M.); voslarovae@vfu.cz (E.V.); bedanovai@vfu.cz (I.B.); vecerekv@vfu.cz (V.V.)

* Correspondence: volfovam@vfu.cz; Tel.: +420-541-562-772

Received: 24 May 2020; Accepted: 8 June 2020; Published: 10 June 2020

Simple Summary: Lemurs kept in captivity are constantly influenced by a number of factors that affect their welfare. All lemur species are classified as endangered species according to the International Union for Conservation of Nature (IUCN) Red List and their population in the wild is decreasing. It is therefore useful to improve the methods for assessing the level of stress in individual lemurs in captivity. In this study, we compared the changes in glucocorticoid concentrations in three species of *Lemuridae* in response to various types of potential stressors during their stay in a temporary housing facility. The glucocorticoid levels were specifically monitored for ring-tailed lemurs (*Lemur catta*), white-headed lemurs (*Eulemur albifrons*) and collared brown lemurs (*Eulemur collaris*).

Abstract: We compared the glucocorticoid concentrations in response to various types of potential stressors present during standard operation of a temporary housing facility between three species, namely, ring-tailed lemurs, collared brown lemurs and white-headed lemurs. The levels of faecal glucocorticoid metabolites (FGMs) were measured non-invasively on a daily basis during a 30-day period. A total of 510 faecal samples were collected. Concentrations of immunoreactive glucocorticoid hormone metabolites were measured in the obtained extracts by using an enzyme immunoassay. The polyclonal antibodies used in this assay were directed against the metabolite 11-oxo-etiocholanolone I. We found all three monitored lemur species to respond to specific potentially stressful situations by increasing ($p < 0.05$) the FGM levels within one to two days after the event. Although housed in the same room, differences in response to potentially stressful situations were found in white-headed lemurs compared to ring-tailed lemurs. Increased mean levels of the FGMs were found more frequently in white-headed lemurs than in ring-tailed lemurs. The results suggest that this species may be more sensitive to changes in its surroundings. In general, the levels of the FGMs showed a similar pattern during 30 days of monitoring suggesting that all groups of lemurs responded in a similar manner to the same events. However, we recorded the differences in the absolute values of glucocorticoid concentrations between the monitored species likely due to the differences in sex ratios in the groups and presence of lactating females in the ring-tailed lemurs.

Keywords: zoo; non-invasive; faecal analysis; glucocorticoid metabolites; enzyme immunoassay

---

1. Introduction

Animals kept in zoos are constantly influenced by a number of factors that affect their welfare. In order to ensure the welfare of captive animals, it is essential to avoid excessive and, in particular, prolonged stress [1]. Even general husbandry tasks, such as handling, transport and veterinary treatment can be potential stressors [2]. These may cause a stress response, especially in more sensitive individuals [2].

Several methods are commonly used to evaluate stress response. They are largely based on the determination of glucocorticoid hormone levels of cortisol and corticosterone or their metabolites [3,4]. A commonly used method for determining the glucocorticoid concentration is to test the blood of the animal. However, in some zoos and in wild animals, the possibility of blood collection is limited and sometimes cannot be performed at all [5,6]. Blood collection brings about many risks, such as injuries caused during catching or sedation of the animal [6]. A more suitable technique is a non-invasive method of assessing stress levels [6], such as the determination of glucocorticoid hormone levels from urine, saliva, fur, milk or animal faeces. This non-invasive assessment of animal welfare is a practical method used for an increasing number of species not only of wild animals, but also of laboratory or farm animals and pets [7]. Collection of faeces seems to be the most appropriate in cases of stress assessment in exotic animals that are not used to handling by humans [8].

Reactions to stressors often vary among different animal species [7,9,10]. Detailed knowledge of the individual species helps to understand their natural behaviour and responses to stressful situations [11]. According to Mason [12], different animal species perceive the conditions of captivity differently. For example, small primates were more stressed by the presence of zoo visitors than larger primates. Differences between ground and tree species were also observed [12]. Weingrill et al. [13] compared the glucocorticoid hormone levels as stress markers in Bornean (*Pongo pygmaeus*) and Sumatran orangutans (*Pongo abelii*). Higher levels of glucocorticoids in the Bornean orangutans were predicted, due to the much more solitary lifestyle and higher frequency of male aggression directed towards females in this species in the wild, which was subsequently confirmed [13]. Furthermore, Cocks [14] found that the hybrids of Bornean and Sumatran orangutans were less resistant to stress than both subspecies, suggesting that the lower survival rate of hybrid orangutans is due to reduced fitness associated with hybridization. Given the variability in species' stress responses and the variety of animals held under human care, understanding species-specific trends in stress responses to daily stressors in captivity can be useful for guiding management practices.

To achieve reliable results via non-invasive monitoring in individual species, knowledge of its metabolism and the mechanism of glucocorticoid hormone excretion is essential [1]. According to Möstl and Palme [1], the concentration of cortisol metabolites in faecal samples reflects the stress-induced glucocorticoid levels with a delay. The species-specific time delay depends on the metabolism and gut passage time of the animal [9]. Steroids are metabolised in the liver and excreted via the bile into the gut. Therefore, measured concentrations of faecal glucocorticoid metabolites (FGMs) reflect an event occurring a certain time ago. This lag time may range from less than 30 min to more than one day, depending on the species and its activity rhythms [5,7,9]. Cambell et al. [15] reported, for instance, that the digestive tract structure varied among different lemurs. They generally have a simple stomach with a slightly prolonged small intestine, sac-like appendix and large intestine of varying length. Black-and-white-ruffed lemurs (*Varecia variegata*) are reported to have a fast passage of the digestive tract (2 to 4 h). Given the very similar structure of the digestive tract in ring-tailed lemurs (*Lemur catta*) [15], the time of intestinal passage will probably be similar. It can be assumed that for similar species belonging to the same family as in this case, after the method has been validated for one species, the same method can be used for a related species, as demonstrated by the following studies evaluating the stress response in lemurs. Balestri et al. [16] compared the faecal glucocorticoid metabolite (FGM) levels among a group of collared brown lemurs (*Eulemur collaris*) living in fragments of forests with a group living in better preserved areas. For the determination of glucocorticoid hormones in the faeces, a successfully validated analysis for this species targeted against the metabolite 11-oxo-etiocholanolone was chosen. Similarly, glucocorticoid levels in faeces during the reproduction season were analysed in red-fronted lemurs (*Eulemur fulvus rufus*) [17] using the analysis targeting the metabolite 11-oxo-etiocholanolone. The suitability of the analysis of the metabolite 11-oxo-etiocholanolone has also been confirmed by studies of the stress level in ring-tailed lemurs and black-and-white-ruffed lemurs by non-invasive methods [18,19].

Lemuridae are generally classified as endangered species according to the International Union for Conservation of Nature (IUCN) Red List and their populations in the wild are decreasing although they are frequently kept in captivity. In order to ensure their stable population, health and welfare in captivity, it is important to reduce the negative effects of stress. A commonly measured endocrine response to stress is the secretion of glucocorticoids [20]. Consequently, there is a need for validation of the methods for assessing the changes in glucocorticoid concentrations in the individual animals non-invasively [1,10].

The aim of this study was to compare the changes in glucocorticoid concentrations in response to various types of potential stressors occurring during a 30-day period in a temporary housing facility in the three selected species of *Lemuridae*, namely ring-tailed lemurs (*Lemur catta*), white-headed lemurs (*Eulemur albifrons*) and collared brown lemurs (*Eulemur collaris*). Given their small body size and fast metabolism, changes in FGMs were expected to be measurable within one to two days after exposure to a stressor [1,10] as confirmed by several studies assessing changes in FGM concentrations resulting from stress, for example, in zoo-living orangutans (*Pongo* spp.) [13], common marmoset (*Callithrix jacchus*), long-tailed macaque (*Macaca fascicularis*), Barbary macaque (*Macaca sylvanus*), chimpanzee (*Pan troglodytes*) and gorilla (*Gorilla gorilla*) [21], Western lowland gorilla (*Gorilla gorilla gorilla*) [22], pileated gibbons (*Hylobates pileatus*) [23], wild gray mouse lemurs [24], ring-tailed lemurs (*Lemur catta*) [18] and black-and-white ruffed lemurs (*Varecia variegata*) [19].

2. Materials and Methods

2.1. Ethics Statement

This study was carried out in strict accordance with the Directive 2010/63/EU of the European Parliament and of the Council of 22 September 2010 on the protection of animals used for scientific purposes and Czech national legislation (i.e., Act no. 246/1992 Coll.) on the protection of animals against cruelty, as amended. All samples were collected non-invasively and animals did not undergo any experimental procedures. Sampling was carried out during temporary housing of three species of *Lemuridae* in an approved holding facility. Since only non-experimental clinical veterinary practices were performed and no handling of animals related to research was carried out, a formal ethics approval from the Animal Welfare Body of the University of Veterinary and Pharmaceutical Sciences Brno with regard to the EU Directive 2010/63/EU was not required.

2.2. Study Subjects and Housing

For the analysis, faecal samples from 9 ring-tailed lemurs, 6 collared brown lemurs and 2 white-headed lemurs were collected for a period of 30 days. The lemurs originated from European zoos and were temporarily housed in a holding facility in Poštovice in the Czech Republic operated by a certified international transportation company, The Nature Resource Network, that was in charge of their transport to other zoos. Details on individual animals monitored in the study are given in Table 1.

All lemurs were fed primate dry formula (pellets), vegetables (carrot, celery, cucumber, lettuce, red beat, onion, boiled rise, tomatoes) and fruits (banana, apple, peach, pineapple, grapes, kiwi) two times in a day, with green leaves, and with cheese and boiled egg once a week. Drinking water was available ad libitum. The lemurs were housed in indoor enclosures (natural light provided from windows) without access to an outdoor area. Changes in the groups during the monitored period are described below.

In ring-tailed lemurs, 9 adult individuals were monitored (8 females and 1 male). There were 4 offspring in the ring-tailed lemurs but they were not included in the analysis. Four adult females (R1–R4) were housed in cage 1 in room 1. Two of them had offspring (R3 had two offspring, R4 had one offspring). The two lactating females (R3 and R4) together with their offspring were separated on day 16 of monitoring to another cage in the same room due to their agonistic behaviour. Female R3 with her offspring was housed in cage 2 and female R4 with her offspring was housed in cage 3.

The remaining two adult females (R1 and R2) were left in cage 1. Subsequently, female R1 was isolated in a separate cage 4 in the same room due to injury on the 23rd day of observation. Another four adult ring-tailed lemurs (R5–R8) were housed in room 2. Each adult was individually housed in separate cages, except for female R6, who was co-housed with her offspring. The first day of monitoring one more adult female ring-tailed lemur (R9) arrived at the facility and was placed in a cage in room 6. Rooms 1 and 6 were interconnected. In white-headed lemurs, 2 individuals (both males, W1 and W2) were housed together throughout the whole monitoring period in room 1 in cage 5. In collared brown lemurs, 6 individuals were assessed: 2 males (C1 and C2) and 4 females (C3–C6). Originally, they were housed together in cage 1 in room 5. During the monitored period, however, they were divided into smaller groups due to mutual attacks and relocated to different cages within the same room (room 5). Namely, on day 16 two males (C1 and C2) were relocated to cage 2, on day 18 one female (C3) was relocated to cage 3, on day 23 one female (C4) was relocated to cage 4 and on day 29 one female (C6) was relocated to cage 5. While the ring-tailed lemurs and white-headed lemurs were housed in the same room, the collared brown lemurs were kept on the other side of the facility. A scheme of the floor plan of the facility and relocation of individuals is provided in Figure 1. All lemurs had visual, olfactory and auditory contact with other animals of the same species during the entire monitored period. All animals were clinically healthy without any signs of health problems or behavioural disorders.

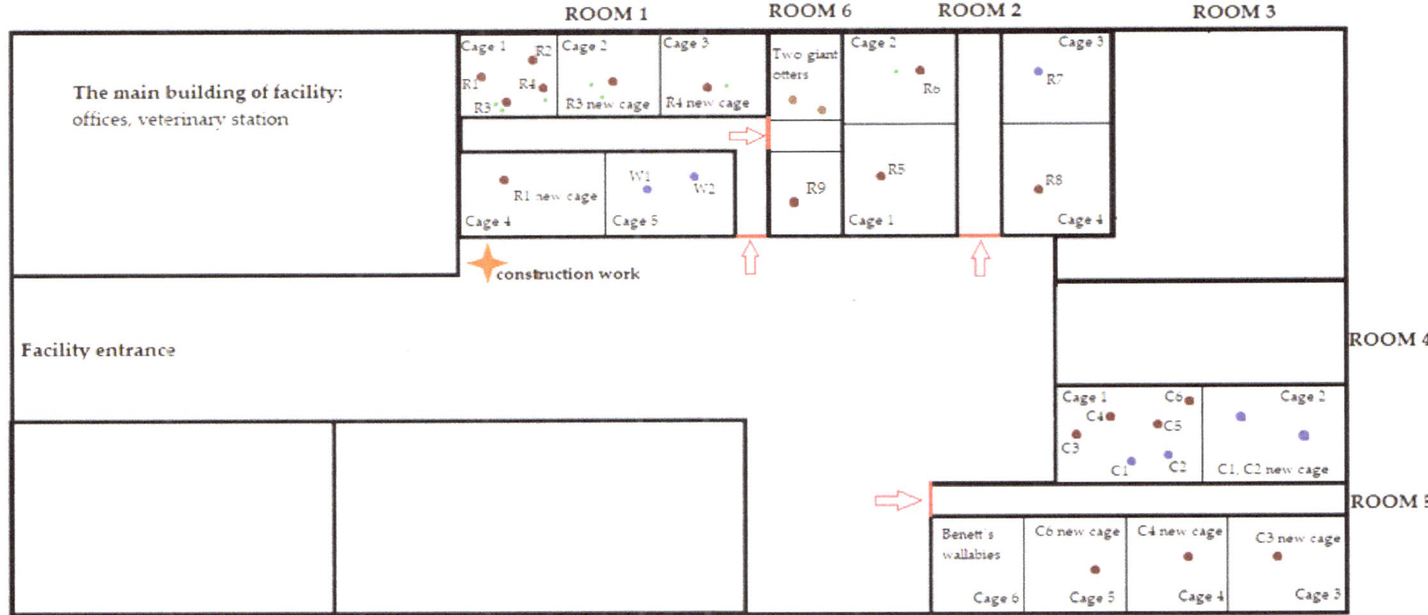

Figure 1. Scheme of floor plan of the facility and relocation of individuals. Red points indicate females; blue points indicate males; green points indicate offspring; red arrows indicate the entrance to the rooms. R1–R9 indicate ring-tailed lemurs 1–9; W1–W2 indicate white-headed lemurs 1–2; C1–C6 indicate collared brown lemurs 1–6.

2.3. Faecal Sampling and Analysis

The monitoring was carried out for 30 days when three lemur species were kept in the facility. Lemurs were exposed to the same husbandry activities (cage cleaning, feeding and watering, visual control of the animals by caretakers). All activities related to the operation of the facility were recorded (e.g., construction work, increased noise, newly incoming animals, transfers within the facility, lactating animals, etc.) by the caretakers during the monitored period in a systematic way with the use of pre-prepared sheets. The faecal samples for the determination of FGMs were collected from the floor once a day between 8:00–12:00 a.m. from each animal, ensuring that they were not contaminated with urine, water or other material. A total of 510 faecal samples were taken. There was no variability in the number of samples taken from different animals. The collected samples were put in plastic sealable bags within 1 h of defecation, labelled and immediately frozen at −20 °C. The subsequent sample processing including extraction was performed at the University of Veterinary Medicine in Vienna, Austria. The samples were thawed, individually homogenised and then successively put into test tubes at weights between 0.480 and 0.520 g. Afterwards 1 mL of distilled water and 4 mL of 100% methanol were added to the weighed samples. Each sample with water and methanol was stirred on a shaker for 30 min and then centrifuged at 4 °C (3750× *g*, 10 min). The detailed extraction process is described in other methodological publications [1,9,25]. Concentrations of immunoreactive glucocorticoid hormone metabolites were measured in the obtained extracts by using an enzyme immunoassay. The polyclonal antibodies used in this assay were directed against the metabolite 11-oxo-etiocholanolone I (laboratory code: 72a, first described by Palme and Möstl [26]), based on previously conducted studies.

2.4. Statistical Analysis

The results were statistically evaluated using the statistical program Unistat 5.6 (Unistat Ltd., London, UK). Data normality was tested using the Shapiro–Wilk test [27]. Since the data were not distributed over the Gaussian curve, the statistical significance of the FGM level differences between the monitored lemur species during the 30 days of observation was determined on each day using a non-parametric Kruskal-Wallis ANOVA test [27] and subsequently nonparametric multiple comparisons with t-distribution using rank sums [27]. For each lemur species, the mean value of the FGMs measured on the first day of the observation was also compared to each subsequent day to determine statistical significance of the differences in the FGM levels by means of the Friedman ANOVA test [27] and subsequently the Dunnett test using rank sums [27]. A value of $p < 0.05$ was considered significant. The first day of sample collection was the day when FGM levels were expected not to be affected by stress, not even in those individuals who arrived on the first day of observation due to a time delay in excretion of FGMs [1]. Therefore, the data from the first day were used as the reference point. FGM levels measured during the monitored period were compared with the timeline of events recorded in the facility in order to determine possible factors eliciting the increase in FGM levels.

3. Results

3.1. Ring-Tailed Lemurs

Records on daily events showed a variety of activities occurring in and outside the rooms where lemurs were kept during a 30-day period of monitoring. In room 1, besides daily routine (e.g., cleaning cages, feeding, visual control of animals), noise from construction work was also recorded. Namely, on the 13th day of the monitored period, all-day construction work was carried out from the other side of the room where the ring-tailed lemurs were housed. The construction work was also performed on day 14. Furthermore, some females originally housed in one group in the same cage were moved to separate cages. On day 16, two females (R3 and R4) with offspring were relocated into two separate cages and on day 23, one female (R1) was moved due to an injury to a separate cage, but all stayed in the same room. On day 29, two giant otters (*Pteronura brasiliensis*) were relocated to room 6 (between room 1 and 2, where ring-tailed lemurs and white-headed lemurs were housed). The lemurs had auditory and olfactory contact with these two otters.

Fluctuations in the mean FGM levels in ring-tailed lemurs (n = 9) over a 30-day period of monitoring are shown in Figure 2 ($F_{(29.232)}$ = 2.3765). On the 14th day of the monitored period, a significant increase ($p < 0.05$) in the mean FGM levels was found. The mean FGM levels also increased on day 17 ($p < 0.01$) and day 24 ($p < 0.05$). Subsequently, on the last day of monitoring, a significant increase in the mean FGM values ($p < 0.01$) was also observed. On other days, the FGM levels did not significantly differ from the first day.

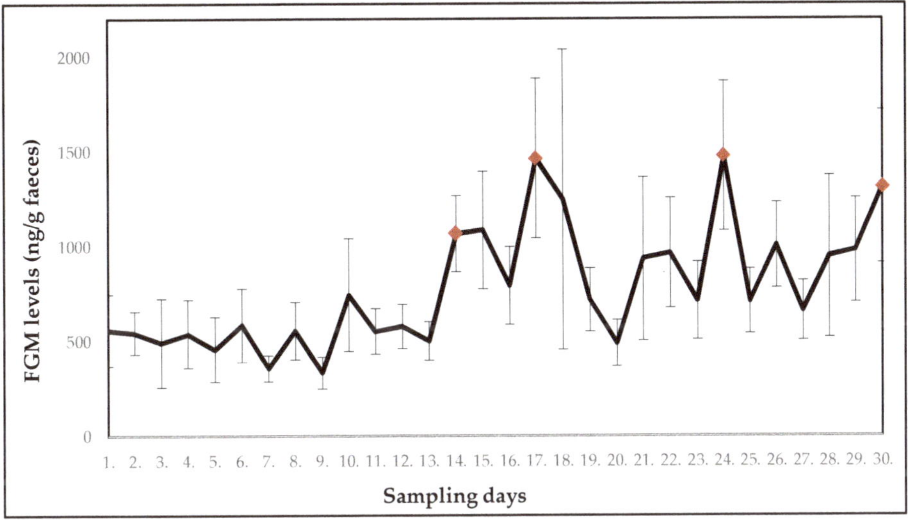

Figure 2. Mean concentrations of the faecal glucocorticoid metabolites (FGMs) in ring-tailed lemurs; significant increases compared to the first day of sampling are indicated. Red points indicate a significant increase in FGM levels.

The mean concentrations of FGMs and potentially stressful events for each day of monitoring are shown in Table 2.

3.2. *White-Headed Lemurs*

Fluctuations in the mean FGM levels in white-headed lemurs (n = 2) housed together in one cage located in the same room as the ring-tailed lemurs over a 30-day period of monitoring are shown in Figure 3 ($F_{(29.29)}$ = 1.6832). A significant increase ($p < 0.05$) in the mean FGM levels was found on days 10, 15, 16, 18, 24, 26, 29 and 30 in comparison with the first day of sampling.

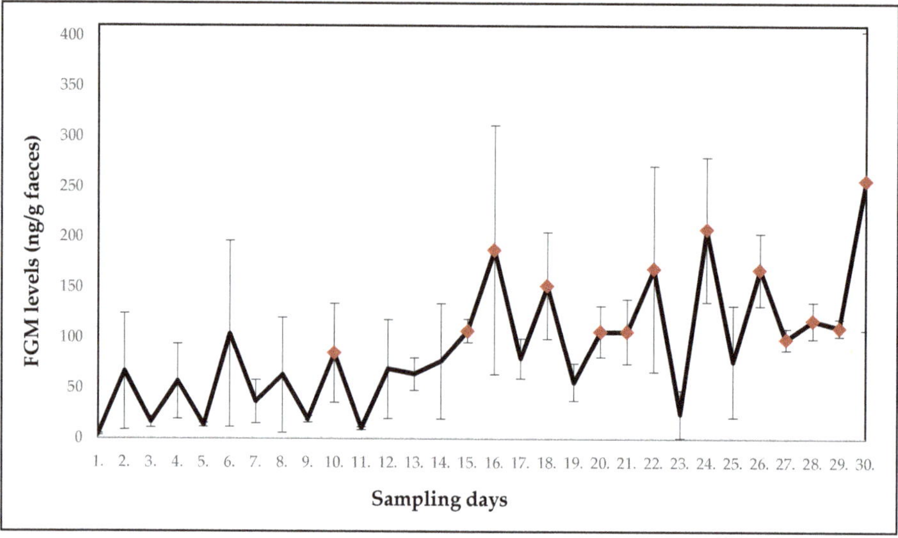

Figure 3. Mean concentrations of the faecal glucocorticoid metabolites (FGMs) in white-headed lemurs; significant increases compared to the first day of sampling are indicated. Red points indicate a significant increase in FGM levels.

Table 3 summarises the mean levels of the FGMs in the white-headed lemurs and potentially stressful events for each day of monitoring.

3.3. Collared Brown Lemurs

They were housed in a different room than the other two lemur species. All of them arrived at the facility on the first day of monitoring and were initially placed together in cage 1 in room 5. During the monitored period, however, some individuals had to be moved due to mutual attacks as described in Section 2. In addition to the changes in the collared brown lemur group, Bennett's wallabies (*Macropus rufogriseus*) arrived on the third day after the beginning of the observation and they were temporarily placed in the same room as the collared brown lemurs.

Fluctuations in the mean FGM levels in collared brown lemurs ($n = 6$) over a 30-day period of monitoring are shown in Figure 4. A statistically significant ($p < 0.05$) increase in the mean FGM levels ($F_{(29.145)} = 2.1247$) was found on days 17, 18, 20, 24 and 30.

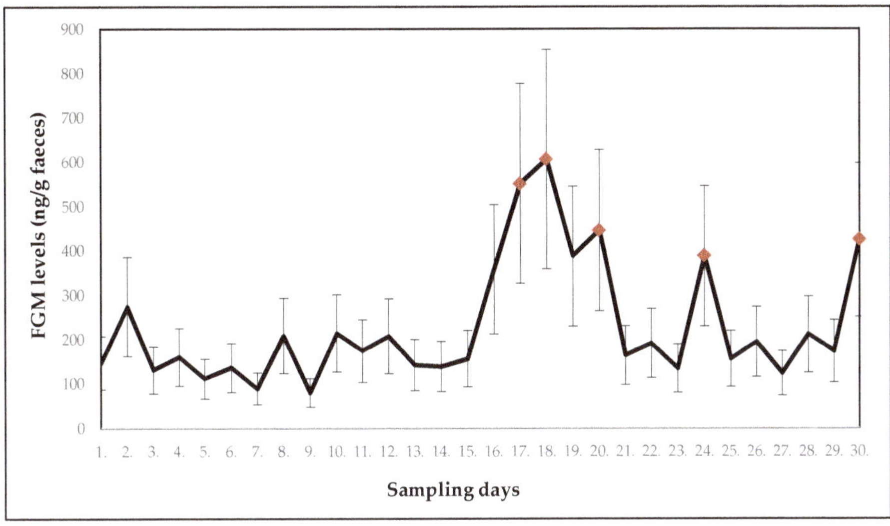

Figure 4. Mean concentrations of the faecal glucocorticoid metabolites (FGMs) in collared brown lemurs with significant increases compared to the first day of sampling indicated. Red points indicate a significant increase in FGM levels.

Table 4 shows the mean FGM levels in collared brown lemurs and potentially stressful events for each day of monitoring.

3.4. Species Comparisons

When comparing the FGM levels between all three lemur species, significant differences ($p < 0.05$) were observed. As shown in Figure 5, the highest levels were measured in ring-tailed lemurs, then in collared brown lemurs, and the lowest concentrations of FGMs were found in white-headed lemurs.

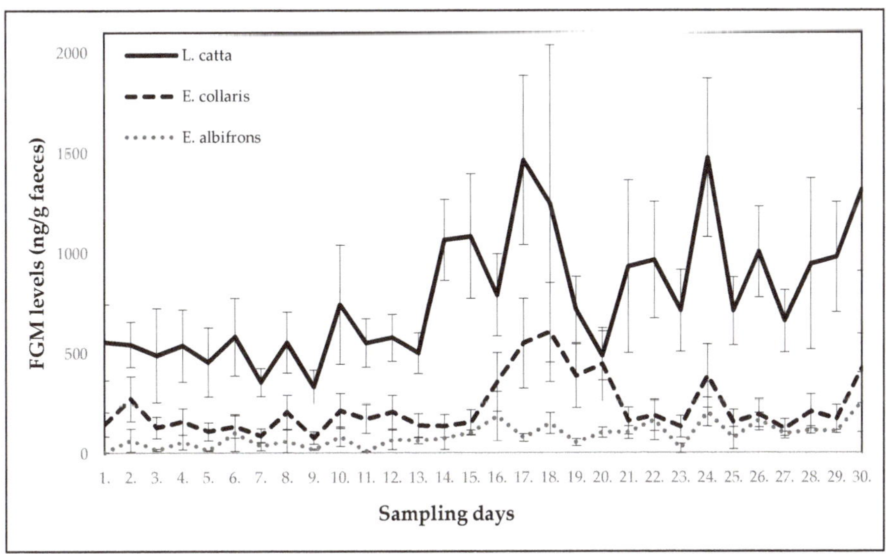

Figure 5. Comparison of glucocorticoid levels in three lemur species.

Table 1. Lemurs observed in a temporary housing facility.

Species	Animal	Sex	Number of Offspring	Birth	Original Zoo	Arrival to the Facility
Ring-tailed lemur	R1	F		unknown	Zoo Wroclaw	1 month before observation
	R2	F		unknown	Zoo Wroclaw	
	R3	F	2	March 2002	Zoo Wroclaw	
	R4	F	1	June 2009	Zoo Wroclaw	
	R5	F		March 2014	zoo Duisburg	3 weeks before observation
	R6	F	1	March 2012	zoo Duisburg	
	R7	F		March 2014	zoo Bernburg	
	R8	M		unknown	Zoo Wroclaw	
	R9	F		March 2016	Hamerton zoo park	first day of observation
White-headed lemur	W1	M		April 2011	Olmense zoo	2 weeks before observation
	W2	M		April 2012	Olmense zoo	
Collared brown lemur	C1	M		April 2015	Hamerton zoo park	first day of observation
	C2	M		April 2015	Hamerton zoo park	
	C3	F		March 2015	Hamerton zoo park	
	C4	F		unknown	Hamerton zoo park	
	C5	F		April 2016	Hamerton zoo park	
	C6	F		April 2015	Hamerton zoo park	

Table 2. Mean concentrations of the faecal glucocorticoid metabolites (FGMs) in ring-tailed lemurs and potentially stressful events for each day of monitoring.

Ring-Tailed Lemur (*Lemur catta*)										
Day	1	2	3	4	5	6	7	8	9	10
Mean ± SE (ng/g faeces)	556.5 ± 189.7	544.3 ± 113.0	490.1 ± 233.8	538.5 ± 180.0	456.1 ± 171.7	582.9 ± 194.2	355.7 ± 69.7	554.2 ± 151.6	332.0 ± 84.7	743.5 ± 297.3
***p*-value**	-	0.099	0.263	0.491	0.646	0.954	0.344	0.774	0.160	0.456
Potential stressor	DR	DR	DR	DR	DR	DR	DR	DR	DR	DR
Day	11	12	13	14	15	16	17	18	19	20
Mean ± SE (ng/g faeces)	552.2 ± 120.3	577.7 ± 116.9	501.1 ± 101.6	1065.9 ± 201.4	1085.2 ± 310.0	792.1 ± 205.2	1464.3 ± 421.9	1246.7 ± 791.8	717.4 ± 166.7	488.0 ± 122.4
***p*-value**	1.000	0.886	0.863	0.037	0.097	0.197	0.006	0.796	0.547	0.752
Potential stressor	DR	DR	construction work	construction work	DR	relocation of two females with offspring	DR	DR	DR	DR
Day	21	22	23	24	25	26	27	28	29	30
Mean ± SE (ng/g faeces)	933.0 ± 430.9	965.5 ± 290.3	712.5 ± 204.8	1476.2 ± 395.2	710.4 ± 170.3	1006.6 ± 225.8	661.3 ± 156.4	946.9 ± 426.6	979.8 ± 275.5	1314.0 ± 403.2
***p*-value**	0.863	0.178	0.863	0.018	0.841	0.136	0.605	0.752	0.109	0.010
Potential stressor	DR	DR	relocation of one female	DR	DR	DR	DR	DR	relocation of giant otters	DR

DR = daily routine

Table 3. Mean concentrations of the faecal glucocorticoid metabolites (FGMs) in white-headed lemurs and potentially stressful events for each day of monitoring.

White-Headed Lemur (*Eulemur albifrons*)										
Day	1	2	3	4	5	6	7	8	9	10
Mean ± SE (ng/g faeces)	3.7 ± 1.0	66.1 ± 57.9	16.1 ± 5.6	56.5 ± 37.3	13.1 ± 2.1	103.5 ± 92.5	36.2 ± 21.6	62.7 ± 57.2	18.8 ± 3.2	84.7 ± 49.2
***p*-value**	-	0.178	0.602	0.246	0.672	0.071	0.301	0.141	0.365	0.041
Potential stressor	DR	DR	DR	DR	DR	DR	DR	DR	DR	DR
Day	11	12	13	14	15	16	17	18	19	20
Mean ± SE (ng/g faeces)	9.8 ± 1.5	68.8 ± 49.2	63.7 ± 16.1	76.6 ± 57.3	106.8 ± 11.7	187.1 ± 123.6	79.3 ± 19.9	151.7 ± 52.9	55.7 ± 18.6	106.2 ± 25.6
***p*-value**	0.649	0.125	0.168	0.054	0.021	0.013	0.086	0.010	0.111	0.017
Potential stressor	DR	DR	construction work	construction work	DR	relocation of three females	DR	DR	DR	DR
Day	21	22	23	24	25	26	27	28	29	30
Mean ± SE (ng/g faeces)	106.2 ± 32.2	168.3 ± 102.3	23.9 ± 23.7	207.5 ± 71.9	76.3 ± 55.7	167.6 ± 35.9	98.7 ± 10.7	117.3 ± 18.3	110.3 ± 8.5	256.0 ± 148.2
***p*-value**	0.016	0.013	0.625	0.002	0.058	0.002	0.033	0.008	0.013	0.005
Potential stressor	DR	DR	relocation of one female	DR	DR	DR	DR	DR	relocation of giant otters	DR

DR = daily routine

Table 4. Mean concentrations of the faecal glucocorticoid metabolites (FGMs) in collared brown lemurs and potentially stressful events for each day of monitoring.

	Collared Brown Lemur (*Eulemur collaris*)									
Day	1	2	3	4	5	6	7	8	9	10
Mean ± SE (ng/g faeces)	146.7 ± 59.9	273.4 ± 111.6	130.6 ± 53.3	159.5 ± 65.1	110.5 ± 45.1	135.4 ± 55.3	88.5 ± 36.1	207.9 ± 84.9	78.8 ± 32.2	213.3 ± 87.1
***p*-value**	-	0.213	0.413	0.803	0.545	0.854	0.393	1.000	0.177	0.721
Potential stressor	DR	DR	relocation of wallabies	DR	DR	DR	DR	DR	DR	DR
Day	11	12	13	14	15	16	17	18	19	20
Mean ± SE (ng/g faeces)	173.6 ± 70.9	206.8 ± 84.4	141.9 ± 57.9	138.5 ± 56.5	156.0 ± 63.7	357.6 ± 146.0	551.1 ± 225.0	606.1 ± 247.4	386.9 ± 158.0	445.5 ± 181.9
***p*-value**	0.803	0.915	0.521	0.830	0.521	0.695	0.010	0.034	0.240	0.047
Potential stressor	DR	DR	DR	DR	DR	relocation of two males	DR	relocation of one female	DR	DR
Day	21	22	23	24	25	26	27	28	29	30
Mean ± SE (ng/g faeces)	163.1 ± 66.6	190.8 ± 77.9	133.9 ± 54.7	387.4 ± 158.2	155.7 ± 63.6	194.0 ± 79.2	122.9 ± 50.2	210.5 ± 85.9	172.8 ± 70.5	424.2 ± 173.2
***p*-value**	0.775	0.498	0.721	0.028	0.873	0.748	0.775	0.618	0.901	0.015
Potential stressor	DR	DR	relocation of one male	DR	DR	DR	DR	DR	relocation of two males and one female	DR

DR = daily routine

4. Discussion

Lemurs in captivity are exposed daily to potentially stressful situations. This is even more pronounced in the case of temporary housing facilities. In addition to the normal daily routine, which includes regular morning cleaning of all quarters and providing food twice a day, construction work was in progress in the monitored facility. Moreover, the observed lemurs were moved, and new animals arrived during the observation period. Such activities that are not carried out regularly are especially likely to cause stress to animals. In all lemur species, significantly increased FGM levels were found during the monitored period. The fluctuation of FGM levels could be caused by the recorded stressors, as elevated concentrations were often found 1 to 2 days after the events that deviated from the daily routine. While blood glucocorticoid concentrations increased shortly after exposure to an acute stressor, and in faeces, the increase of glucocorticoids resulting from stress was seen with a delay [7].

On the 13th and 14th day of the monitored period, all-day construction work was carried out near the room where the ring-tailed lemurs were housed. In the days after (14th and 15th day) a significant increase ($p < 0.05$) in the mean FGM levels was found. This increase could result from ongoing construction work as lemurs were exposed to increased noise and vibrations in their cages. The mean FGM levels also increased on day 17 ($p < 0.01$) and day 24 ($p < 0.05$). These increases could be related to relocations of two females with offspring (R3 and R4) into separate cages (16th and 23rd day). The separation was necessary due to agonistic behaviour observed in the females. Despite the females being familiar with each other from their original location, stress from the new environment could cause changes in their behaviour. Relocation can be perceived as a stressor in captive wild animals because they perceive all changes and new situations as a potential threat [28]. On day 23, one female (R1) was moved to a separate cage in the same room due to an injury. The injured female exhibited no significant increase in FGM levels, however, mean FGM levels in the group of ring-tailed lemurs increased. Finally, also on the last day of monitoring, a significant increase in the mean FGM values ($p < 0.01$) was observed. According to records, two giant otters (*Pteronura brasiliensis*) were relocated into the facility on the previous day which could be stressful for lemurs housed nearby. In the monitored period, relatively stable FGM levels were found on days which were not preceded by any exceptional events (daily routine was followed). Assumedly, the above-mentioned disturbances were responsible for the changes in FGM levels and they showed a consistent one-day delay before elevating FGM concentrations. White-headed lemurs were exposed to the same potentially stressful events as the ring-tailed lemurs. Construction work was in progress on day 13 and 14 near their enclosure, which could explain a significant increase ($p < 0.05$) in the mean FGM levels on days 15 and 16. Interestingly, there was a two-day delay before the FGMs elevated in white-headed lemurs in contrast with the one-day delay in ring-tailed lemurs following the same stressor. The significantly elevated levels of FGMs were found also in the last days of monitoring (day 26 to 29). The reason is unclear. The records did not show any deviations from the daily routine. However, the animals might have responded even to some minor changes not detected or perceived by the caretakers. According to Morgan and Tromborg [2], animals can respond to the inability to escape from human reach and the daily routine of cleaning the enclosure. Although routine is a repeated activity for a human being, an animal in captivity may perceive this situation differently and may be stressed. For example, primates have been shown to have high heart rates at the time of the keeper's arrival at the enclosure due to routine cleaning [2]. Thus, for some more sensitive individuals, the presence of a person in their vicinity can be stressful, especially if they are in an unknown place [2,29].

In both, the ring-tailed lemurs and the white-headed lemurs, a significant ($p < 0.05$) increase in FGM levels was recorded following the construction work throughout the day in the yard of the facility where both species were housed. As Morgan and Tromborg [2] reported, excessive noise caused by, for example, noisy construction work, is stressful and affects not only the animal's behaviour but also the levels of specific stress markers. In captive wild animals, acute and chronic stress caused by difficulties in coping with stressors such as public presence and noise, among others, can induce a significant increase in FGMs and repetitive pathological behaviours such as stereotypies [30,31].

Collared brown lemurs were housed in a different room than the other two lemur species. Thus, they could be affected by different potentially stressful events. In particular, these included necessary relocations due to the occurrence of agonistic behaviour in the group. In the monitored period, a significant increase ($p < 0.05$) in the mean FGM levels was found on days 17, 18, 20, 24 and 30, which could be caused by aggression and subsequent relocations. Mutual fights escalated especially during feeding. Aggressive behaviour has been proven to affect FGM levels, with dominant individuals having higher glucocorticoid levels [32–34]. However, also the movement itself or other changes in the group affect the level of the FGMs, as was demonstrated in the ring-tailed lemurs.

While the significant increase of FGMs in ring-tailed and white-headed lemurs was caused the next day after the arrival of giant otters, in collared brown lemurs no significant change was found in FGM levels following the arrival of Bennett's wallabies (*Macropus rufogriseus*) to the same room. Furthermore, in collared brown lemurs there was no increase in the mean FGM levels in response to the construction work carried out on the 13th and 14th day of the observation, likely because the collared brown lemurs were located on the other side of the facility, thus no noises or vibrations from the construction work were recorded here.

During the monitored period, handling of animals was most often associated with their relocation to other cages due to agonistic behaviour. For relocation, capture and direct handling was necessary. That can be perceived as a strong stressor [24,28] with handling duration affecting the stress response and subsequently FGM levels [35]. However, it also depends on other factors, such as temperament of the individual and previous experience with the stressor. Considering their activity, it is generally recommended to capture lemurs in early morning hours and only by experienced caretakers [36,37]. The lemurs who were relocated in our study were captured by an experienced caretaker with the use of net and transferred to the new cage. Capture of animals is considered an acute stressor, but it can have longer lasting effects if it is carried out repeatedly in a short period [38]. According to Balcombe et al. [38], handling of animals is one of the most common husbandry practices but it results in variously strong stress responses and affects the activity of the immune system of animals. Hämäläinen et al. [24] measured the level of stress in grey mouse lemurs (*Microcebus murinus*) based on the determination of elevated levels of FGMs after capture. Although elevated levels have been recorded in some cases, routine capture has not had long-term consequences. The stress response in black howler monkeys (*Alouatta caraya*) was evaluated by Rangel-Negrín et al. [39]. The results showed that after capture, the levels of FGMs were affected. Similarly, Volfova et al. [18] recorded the increase of FGMs in response to handling in ring-tailed lemurs. Captive wild animals are expected to have a stronger response to handling stress than domesticated animals [40]. Given that handling is one of the primary stressors for captive wild animals, it is essential to monitor its potential negative impact on their welfare and avoid it as much as possible.

The results show that all the three monitored lemur species respond during the observed period to specific stressful situations by increasing ($p < 0.05$) the FGM levels similarly within one to two days after the event. This delay from the stress event to GC excretion into faeces was confirmed not only in previous studies regarding handling of the ring-tailed lemurs and transportation of the black-and-white ruffed lemurs, but also in many other studies concerning primates [13,21,23,24,41]. Generally, the delay is 24 to 48 h in primates [22]. In white-headed lemurs, however, differences in response to potentially stressful situations were reported compared to ring-tailed lemurs. Increased average levels of the FGMs were found more frequently in white-headed lemurs than in ring-tailed lemurs. The results may indicate that this species responds more sensitively to changes in its surroundings. Nevertheless, these differences may also be due to the different temperament of the animals, as some studies confirm [6,7], as well as by the differences in the structure of the gastrointestinal tract in different lemurs species [7,42], which affects the amount of glucocorticoids secreted into the faeces [6]. It has been proven that there are differences in the metabolism of glucocorticoid hormones and their secretion into faeces between different species [6,43]. However, closely related species can be expected to have a similar or even equal representation of glucocorticoid hormone metabolites in their faeces and the time for which there is a demonstrable increase in the FGMs.

The differences in the FGM concentrations in the individual species may be also due to different proportions of sex in the observed groups [44], because sex has a demonstrable impact on glucocorticoid levels, with females in oestrous, pregnant or lactating exhibiting elevated circulating glucocorticoids [6]. In our case, the ring-tailed lemur group consisted of eight females and one male, which is the standard group structure even in the wild. Three of these females had offspring and were lactating; the GC levels measured in their faeces thus increased the mean FGMs of the whole group. Similarly, Starling et al. [45] found elevated levels of the FGMs in lactating ring-tailed lemur females living in the wild. The group of white-headed lemurs, where the mean FGM levels were the lowest, consisted of two males. The group of collared brown lemurs consisted of four females and two males. The influence of sex on FGM levels was also confirmed by Arias et al. [46], who reported higher cortisol levels in female lama guanicoe (*Guanaco*) compared to males due to pregnancy. Dantzer et al. [47] found female American red squirrel (*Tamiasciurus hudsonicus*) also have elevated levels of FGMs during pregnancy. Furthermore, Carrera et al. [48] found higher glucocorticoid levels to be associated with low rank (compared to high rank) and first-time mothers (compared to multiparous mothers) in geladas (*Theropithecus gelada*).

The study has shown that some situations and activities related to captive housing can result in elevated levels of FGMs in lemurs. A similar pattern of changes in the FGM levels in response to potentially stressful situations was observed in the monitored lemurs. All three lemur species in our case showed a significant increase ($p < 0.05$) in the FGM levels within one, or at most two days after the stressor exposure. Total concentrations of the FGMs differed between the species, with the highest in ring-tailed lemurs, lower in collared brown lemurs and the lowest in white-headed lemurs. In white-headed lemurs, much lower FGM levels were found in comparison with the other two lemur species, which could be interpreted as a species difference. However, it is also possible that their low levels of FGMs corresponded to the fact that as the only one their group was not divided during the monitored period, and the males were used to each other. For social species, their mutual dependence is high [49]. Furthermore, it was also the only male group. The absence of females could also lead to an overall lower level of FGMs. In contrast, the highest glucocorticoid concentrations in ring-tailed lemurs could have been caused by the presence of lactating females in the group. Furthermore, the individual animal's temperament or social status may affect the absolute FGM levels.

Limitations of the study included lack of baseline data, varying time of arrival of individual animals to the facility (and thus varying period of acclimatization or lack thereof before the commencement of the study), different number of animals and varying sex ratio in the monitored groups. However, it was not possible to experimentally design and control the structure of the monitored groups and the conditions of observation.

5. Conclusions

Stress is currently one of the most serious problems in rearing endangered animal species in zoos and other facilities around the world. For their preservation and successful breeding, it is therefore of the utmost importance to avoid excessive stress and to provide the most suitable conditions for their life in captivity. In order for the housing conditions to be properly assessed, it is necessary to determine the impact of the factors to which captive lemurs are exposed. This study shows fluctuations in FGM levels in three lemur species likely resulting from changes in the structure of a group with relocation of individuals, construction work in the immediate vicinity of animals or placing another animal in quarters adjacent to the observed individuals. The results suggest that some lemur species may be more or less sensitive to such disturbances. However, FGM concentrations had a similar pattern (corresponding to the occurrence of potentially stressful events) during 30 days of monitoring in all three species of *Lemuridae* suggesting that closely related animal species respond to stress load similarly. A significant increase in the FGM levels was found within one, or at most two days after the exposure to a stressor.

Author Contributions: Conceptualization, M.V. and Z.M.; data curation, M.V. and I.B.; methodology, M.V. and Z.M.; supervision, E.V. and V.V.; writing—original draft, M.V.; writing—review and editing, E.V., I.B., and V.V. All authors have read and agreed to the published version of the manuscript.

Funding: This research received no external funding.

Acknowledgments: We wish to acknowledge the advice of Rupert Palme, of the University of Veterinary Medicine, Vienna. Furthermore, we would like to thank the Nature Resource Network facility operators in the Czech Republic, namely Bojana Bobić Gavrilović and Miroslav Poljak, for making the research possible and for their help with sample collection, and Edith Klobetz Rassam and Elke Leitner for technical assistance with the EIA analysis.

Conflicts of Interest: The authors declare no conflicts of interest.

References

1. Möstl, E.; Palme, R. Hormones as indicators of stress. *Domest. Anim. Endocrinol.* **2002**, *23*, 67–74. [CrossRef]
2. Morgan, K.N.; Tromborg, C.H.T. Sources of stress in captivity. *Appl. Anim. Behav. Sci.* **2007**, *102*, 262–302. [CrossRef]
3. Pribbenow, S.; Jewgenow, K.; Vargas, A.; Rodrigo, S.; Naidenko, S.; Dehnhard, M. Validation of an enzyme immunoassay for the measurement of faecal glucocorticoid metabolites in Eurasian (*Lynx lynx*) and Iberian lynx (*Lynx pardinus*). *Gen. Comp. Endocrinol.* **2014**, *206*, 166–177. [CrossRef] [PubMed]
4. Bashaw, M.J.; Sicks, F.; Palme, R.; Schwarzenberger, F.; Tordiffe, A.S.W.; Ganswindt, A. Non-invasive assessment of adrenocortical activity as a measure of stress in giraffe (*Giraffa camelopardalis*). *BMC Vet. Res.* **2016**, *235*, 1–13. [CrossRef] [PubMed]
5. Palme, R.; Rettenbacher, S.; Touma, C.; El-Bahr, S.M.; Möstl, E. Stress hormones in mammals and birds: Comparative aspects regarding metabolism, excretion, and noninvasive measurement in fecal samples. *Ann. NY Acad. Sci.* **2005**, *1040*, 162–171. [CrossRef] [PubMed]
6. Palme, R. Non-invasive measurement of glucocorticoids: Advances and problems. *Physiol. Behav.* **2019**, *199*, 229–243. [CrossRef] [PubMed]
7. Touma, C.; Palme, R. Measuring fecal glucocorticoid metabolites in mammals and birds: The importance of validation. *Ann. NY Acad. Sci.* **2005**, *1046*, 54–74. [CrossRef] [PubMed]
8. Touma, C.; Palme, R.; Sachser, N. Analyzing corticosterone metabolites in fecal samples of mice: A noninvasive technique to monitor stress hormones. *Horm. Behav.* **2004**, *45*, 10–22. [CrossRef] [PubMed]
9. Palme, R. Measuring fecal steroids: Guidelines for practical application. *Ann. NY Acad. Sci.* **2005**, *1046*, 75–80. [CrossRef] [PubMed]
10. Millspaugh, J.J.; Washburn, B.E. Use of fecal glucocorticoid metabolite measures in conservation biology research: Considerations for application and interpretation. *Gen. Comp. Endocrinol.* **2004**, *138*, 189–199. [CrossRef] [PubMed]
11. Mason, G.; Mendl, M. Why is there no simple way of measuring animal welfare? *Anim. Welf.* **1993**, *2*, 301–319.
12. Mason, G.J. Species differences in responses to captivity: Stress, welfare and the comparative method. *Trends Ecol. Evol.* **2010**, *25*, 713–721. [CrossRef] [PubMed]
13. Weingrill, T.; Willems, E.P.; Zimmermann, N.; Steinmetz, H.; Heistermann, M.M. Species-specific patterns in fecal glucocorticoid and androgen levels in zoo-living orangutans (*Pongo* spp.). *Gen. Comp. Endocrinol.* **2011**, *172*, 446–457. [CrossRef] [PubMed]
14. Cocks, L. Factors affecting mortality, fertility, and well-being in relation to species differences in captive orangutans. *Int. J. Primatol.* **2007**, *28*, 421–428. [CrossRef]
15. Campbell, J.L.; Eisemann, J.H.; Williams, C.V.; Glenn, K.M. Description of the gastrointestinal tract of five lemur species: *Propithecus tattersalli, Propithecus verreauxi coquereli, Varecia variegata, Hapalemur griseus*, and *Lemur catta*. *Am. J. Primatol.* **2000**, *52*, 133–142. [CrossRef]
16. Balestri, M.; Barresi, M.; Campera, M.; Serra, V.; Ramanamanjato, J.B.; Heistermann, M.; Donati, G. Habitat degradation and seasonality affect physiological stress levels of *Eulemur collaris* in littoral forest fragments. *PLoS ONE* **2014**, *9*, e107698. [CrossRef] [PubMed]
17. Ostner, J.; Kappeler, P.; Heistermann, M. Androgen and glucocorticoid levels reflex seasonally occurring social challenges in male red fronted lemurs (*Eulemur fulvus rufus*). *Behav. Ecol. Sociobiol.* **2008**, *62*, 627–638. [CrossRef] [PubMed]

18. Volfova, M.; Machovcova, Z.; Schwarzenberger, F.; Voslarova, E.; Bedanova, I.; Vecerek, V. Non-invasive assessment of adrenocortical activity as a measure of stress in ring-tailed lemurs (*Lemur catta*). *Berl. Munch. Tierarztl. Wochenschr.* **2020**, *133*, 28–35.
19. Volfova, M.; Machovcova, Z.; Schwarzenberger, F.; Voslarova, E.; Bedanova, I.; Vecerek, V. The effects of transport stress on the behaviour and adrenocortical activity of the black-and-white ruffed lemur (*Varecia variegata*). *Acta Vet. Brno* **2019**, *88*, 85–92. [CrossRef]
20. Cockrem, J.F. Individual Variation in Glucocorticoid Stress Responses in Animals. *Gen. Comp. Endocrinol.* **2013**, *181*, 45–58. [CrossRef] [PubMed]
21. Heistermann, M.; Palme, R.; Ganswindt, A. Comparison of different enzyme immunoassays for assessment of adrenocortical activity in primates based on fecal analysis. *Am. J. Primatol.* **2006**, *68*, 257–273. [CrossRef] [PubMed]
22. Clark, F.E.; Fitzpatrick, M.; Hartley, A.; King, A.J.; Lee, T.; Routh, A.; Walker, S.L.; George, K. Relationship between behavior, adrenal activity, and environment in zoo-housed Western lowland gorillas (*Gorilla gorilla gorilla*). *Zoo Biol.* **2011**, *30*, 1–16. [CrossRef] [PubMed]
23. Pirovino, M.; Heistermann, M.; Zimmermann, N.; Zingg, R.; Clauss, M.; Codron, D.; Kauphanspeter, F.J.; Steinmetz, W. Fecal glucocorticoid measurements and their relation to rearing, behavior, and environmental factors in the population of pileated gibbons (*Hylobates pileatus*) held in European zoos. *Int. J. Primatol.* **2011**, *32*, 1161–1178. [CrossRef]
24. Hämäläinen, A.; Heistermann, M.; Fenosoa, Z.S.E.; Raus, C. Evaluating capture stress in wild gray mouse lemurs via repeated fecal sampling: Method validation and the influence of prior experience and handling protocols on stress responses. *Gen. Comp. Endocrinol.* **2014**, *195*, 68–79. [CrossRef] [PubMed]
25. Palme, R.; Touma, C.; Arias, N.; Dominchin, M.F.; Lepschy, M. Steroid extraction: Get the best out of faecal samples. *Wien. Tierarztl. Monatsschr.* **2013**, *100*, 238–246.
26. Palme, R.; Möstl, E. Measurement of cortisol metabolites in faeces of sheep as a parameter of cortisol concentration in blood. *Mamm. Biol.* **1997**, *62*, 192–197.
27. Zar, J.H. *Biostatistical Analysis*, 4th ed.; Prentice-hall, INC: New Jersey, NJ, USA, 1999.
28. Grandin, T. Assessment of stress during handling and transport. *Sci. J. Anim. Sci.* **1997**, *75*, 249–257. [CrossRef] [PubMed]
29. Margulis, S.W.; Hoyos, C.; Anderson, M. Effect of felid activity on zoo visitor interest. *Zoo Biol.* **2003**, *22*, 587–599. [CrossRef]
30. Romano, M.C.; Rodas, A.Z.; Valdez, R.A.; Hernández, S.E.; Galindo, F.; Canale, D.; Brousset, D.M. Stress in wildlife species: Noninvasive monitoring of glucocorticoids. *Neuroimmunomodulation* **2010**, *17*, 209–212. [CrossRef] [PubMed]
31. Owen, M.A.; Swaisgood, R.R.; Czekala, N.M.; Steinman, K.; Lindburg, D.G. Monitoring stress in captive giant pandas (*Ailuropoda melanoleuca*): Behavioral and hormonal responses to ambient noise. *Zoo Biol.* **2004**, *23*, 147–164. [CrossRef]
32. Cavigelli, S.A. Behavioural patterns associated with faecal cortisol levels in free rating female Ring-tailed lemurs, *Lemur catta*. *Anim. Behav.* **1999**, *57*, 935–944. [CrossRef] [PubMed]
33. Cavigelli, S.A.; Dubovick, T.; Levash, W.; Jolly, A.; Pitts, A. Female dominance status and fecal corticoids in a kooperative breeder with low reproductive skew: Ring-Tailed lemurs (*Lemur catta*). *Horm. Behav.* **2003**, *43*, 166–179. [CrossRef]
34. Gould, L.; Ziegler, T.E.; Wittwer, D.J. Effects of reproductive and social variables on fecal glucocorticoid levels in a sample of adult male ring-tailed lemurs (*Lemur catta*) at the Beza Nahafaly Reserve, Madagascar. *Am. J. Primatol.* **2005**, *67*, 5–23. [CrossRef] [PubMed]
35. Bennett, K.A.; Moss, S.E.W.; Pomeroy, P.; Speakman, J.R.; Fedak, M.A. Effects of handling regime and sex on changes in cortisol, thyroid hormones and body mass in fasting grey seal pups. *Comp. Biochem. Phys. A* **2012**, *161*, 69–76. [CrossRef] [PubMed]
36. McGillivray, C.L. *Husbandry Manual for Black and Ehite Ruffed Lemur—Varecia variegata variegata (Mammalia: Lemuridae)*; Western Sydney Institute of Tafe: Richmond, Australia, 2007.
37. Quine, H. *Husbandry Guidelines for White Fronted Brown Lemur (Eulemur fulvus albifrons)*; Western Sydney Institute of Tafe: Richmond, Australia, 2009.
38. Balcombe, J.P.; Barnard, N.D.; Sandusky, C. Laboratory routines cause animal stress. *J. Am. Assoc. Lab. Anim. Sci.* **2004**, *43*, 42–51.

39. Rangel-Negrín, A.; Flores-Escobar, E.; Chavira, R.; Canales-Espinosa, D.; Dias, P.A.D. Physiological and analytical validations of fecal steroid hormone measures in black howler monkeys. *Primates* **2014**, *55*, 459–465. [CrossRef] [PubMed]
40. Snyder, R.J.; Perdue, B.M.; Powell, D.M.; Forthman, D.L.; Bloomsmith, M.A.; Maple, T.L. Behavioral and hormonal consequences of transporting giant pandas from China to the United states. *J. Appl. Anim. Welf. Sci.* **2012**, *15*, 1–20. [CrossRef] [PubMed]
41. Shutt, K.; Setchell, J.M.; Heistermann, M. Non-invasive monitoring of physiological stress in the estern lowland gorila (*Gorilla gorilla gorilla*): Validation of fecal glucocorticoid assay and methods for practical application in the field. *Gen. Comp. Endocrinol.* **2012**, *179*, 167–177. [CrossRef] [PubMed]
42. Boccia, M.L.; Laudenslager, M.L.; Reite, M.L. Individual differences in Macaques' responses to stressors based on social and physiological factors: Implications for primate welfare and research outcomes. *Lab. Anim.* **1995**, *29*, 250–257. [CrossRef] [PubMed]
43. Wasser, S.K.; Hunt, K.E.; Brown, J.L.; Cooper, K.; Crockett, C.M.; Bechert, U.; Millspaugh, J.J.; Larson, S.; Monfort, S.L. A generalized fecal glucocorticoid assay for use in a diverse array of nondomestic mammalian and avian species. *Gen. Comp. Endocrinol.* **2000**, *120*, 260–275. [CrossRef] [PubMed]
44. Cavigelli, S.A.; Caruso, M.J. Sex, social status and physiological stress in primates: The importace of social and glucocorticoid dynamics. *Philos. Trans. R. Soc.* **2015**, *370*, 1–13. [CrossRef] [PubMed]
45. Starling, A.P.; Charpentier, M.J.E.; Fitzpatrick, C.; Scordato, E.S.; Drea, C.H.M. Seasonality, sociality, and reproduction: Long-Term stressors of ring-tailed lemurs (*Lemur catta*). *Horm. Behav.* **2010**, *57*, 76–85. [CrossRef] [PubMed]
46. Arias, N.; Requena, M.; Palme, R. Measuring faecal glucocorticoid metabolites as a non-invasive tool for monitoring adrenocortical activity in South american camelids. *Anim. Welf.* **2013**, *22*, 25–31. [CrossRef]
47. Dantzer, B.; McAdam, A.G.; Palme, R.; Fletcher, Q.E.; Boutin, S.; Humphries, M.M.; Boonstra, R. Fecal cortisol metabolite levels in free-ranging North American red squirrels: Assay validation and the effects of reproductive condition. *Gen. Comp. Endocrinol.* **2010**, *167*, 279–286. [CrossRef] [PubMed]
48. Carrera, S.C.; Sen, S.; Heistermann, M.; Lu, A.; Beehner, J.C. Low rank and primiparity increase fecal glucocorticoid metabolites across gestation in wild geladas. *Gen. Comp. Endocrinol.* **2020**, *293*, 113494. [CrossRef] [PubMed]
49. MacLean, E.L.; Sandel, A.A.; Bray, J.; Oldenkamp, R.E.; Reddy, R.B.; Hare, B.A. Group Size Predicts Social but Not Nonsocial Cognition in Lemurs. *PLoS ONE* **2013**, *8*, e66359. [CrossRef] [PubMed]

Article

Physiological Stress Reactions in Red Deer Induced by Hunting Activities

Sofia Vilela [1], António Alves da Silva [1], Rupert Palme [2], Kathreen E. Ruckstuhl [3], José Paulo Sousa [1] and Joana Alves [1,*]

1 Centre for Functional Ecology (CFE), Department of Life Sciences, University of Coimbra, Calçada Martim de Freitas, 3000-456 Coimbra, Portugal; vpsofia.a@gmail.com (S.V.); antonioalvesdasilva@gmail.com (A.A.d.S); jps@zoo.uc.pt (J.P.S.)
2 Unit of Physiology, Pathophysiology and Experimental Endocrinology, Department of Biomedical Sciences, University of Veterinary Medicine, Veterinärplatz 1, Vienna 2210, Austria; rupert.palme@vetmeduni.ac.at
3 Department of Biological Sciences, University of Calgary, 2500 University Drive NW, Calgary, AB T2N 1N4, Canada; kruckstu@ucalgary.ca
* Correspondence: joanasilvaalves@gmail.com

Received: 19 May 2020; Accepted: 5 June 2020; Published: 8 June 2020

Simple Summary: Game hunting is an activity largely practiced all over the world. Understanding its consequences on wildlife is crucial for the proper management and development of hunting directives. In this study, we examined stress levels in hunted wild red deer by assessing cortisol levels and its metabolites in multi-temporal biological samples. Overall, we found evidence for an influence on stress levels of red deer caused by repeated exposure to hunting events, which could have important implications on the sustainability and conservation of wild populations. Furthermore, our results highlight the use of hair samples as a useful long-term stress indicator.

Abstract: Hunting activity is usually seen as a factor capable of causing an intense stress response in wildlife that may lead to short but also long-term stress. In the Lousã Mountain, Portugal, the population of red deer (*Cervus elaphus*) is the target of intensive seasonal hunting. We collected and measured cortisol (and its metabolites) in three tissues types (blood, feces and hair) from red deer hunted during two hunting seasons to evaluate the stress levels at different time windows. We also assessed the immunological and physical condition of the animals. We predicted that the hunting activity would act as a stressor inducing increased short and long-term stress levels in the population. Results showed an increase in hair cortisol levels during the months of harvesting. Surprisingly, the tendency for plasma cortisol levels was to decrease during the hunting season, which could be interpreted as habituation to hunting activity, or due to the hunting duration. Contrary to our predictions, fecal cortisol metabolites did not show any clear patterns across the months. Overall, our results suggest an influence of hunting activities on the physiological stress in red deer. In addition, hair seems to be useful to measure physiological stress, although more studies are required to fully understand its suitability as an indicator of long-term stress. Methodologically, our approach highlights the importance of simultaneously using different methods to assess short and long-term effects in studies on physiological stress reactions.

Keywords: *Cervus elaphus*; plasma; feces; hair; glucocorticoids; hunting; stress

1. Introduction

Stress responses occur when an animal perceives an external noxious stimulus (stressor) such as predation, adverse weather, habitat change, or anthropogenic disturbances. In physiological terms, such stress reactions can be described as a cascade of endocrine secretions involving the hypothalamic-pituitary-adrenocortical axis (HPA) and sympatho-adrenomedullary system (SAS), wherein the adrenal glands play an important role [1,2]. These physiological mechanisms are induced to avoid, survive, or recover from an adverse condition [3], and involve the increase of glucocorticoid and/or catecholamine secretion leading to cumulative costs defined as allostatic load. When energy requirements of the organism exceed energy reserves, allostatic overload (type 1) happens [4]. Once the threshold is reached, the emergency response can be triggered, redirecting the animal towards survival and interrupting its normal life history. At a short-term scale, the rise of glucocorticoids (GC) induces physiological effects such as the regulation of the immune system, the increase in glucose created from the body's energy stores, and contributes to memory consolidation [3,5,6]. However, long-term exposure to stressors may trigger type 2 allostatic overload, known as chronic stress [4]. At this point, there is sufficient or even excess energy consumption which can have negative consequences such as hypertension, inhibition of the immune system, promotion of neuronal cell death and reduction of body growth [3,5,6]. Additionally, with chronic, non-lethal stressors, animals can exhibit habituation to the stressor, which means that after a certain period of exposure the animals no longer perceive the stimulus as noxious. Therefore, at this point, the rise of GC is no longer triggered [2].

The quantification of glucocorticoids is an important tool to study the neuro-endocrine stress axis and can provide insights into an animal's well-being as well as a better understanding of ecological and evolutionary processes, which can be useful for conservation purposes [7–12]. The determination of cortisol or corticosterone in plasma is widely used as an indicator of stress [1,12]. This measure represents the instant stress status of an animal and can be used as a tool for experimental studies to understand the effects of specific stressors such as predators or weather conditions [12–14]. However, since this method requires the handling of animals for blood collection, which in itself represents a stressful situation for animals, new methods with non-invasive sampling have been developed. Above all, analysis of glucocorticoid metabolites in feces has been established as a validated, successful alternative to blood, in several studies [1,9,10,15,16]. In red deer, using fecal samples, maximum levels of cortisol metabolites are usually recorded with a delay of approximately 18 h after the disturbance [17]. In addition, cortisol in hair has been used as an indicator of stressful conditions [18,19]. Some recent studies suggest that hair shows levels of GC that are accumulated for some weeks to months, depending on the species [20–23].

It has been shown that hunting is capable of causing an intense stress response in wildlife, which can significantly affect animal welfare. Bateson and Bradshaw [24] showed that hunting triggers physiological effects in red deer (*Cervus elaphus*), such as disruption of muscle tissue and alterations of concentrations of β-endorphin. Moreover, the same authors demonstrated that cortisol levels increased considerably when animals were exposed to this kind of harvesting. Although its effects are poorly explored, hunting activities, due to their seasonality, can result in repeated and prolonged stress. Recently, a study suggested that active hunting events using hounds or involving great densities of people can have a stronger impact with higher cortisol levels in cervids than calmer stalking [25].

In Portugal and Spain, the main hunting process used is a non-selective process called "montaria", which involves the release of packs of dogs (usually 150 to 300 dogs), that will chase the animals during a maximum of five hours, and move them toward the hunters, positioned at the edge of the hunting area [26,27]. Red deer is an important game species in Portugal, and its exploitation is mostly done with this type of hunting process [26]. Due to the effects that this hunting practice can have on wildlife, its ethical and economical effects have been debated [28]. Notwithstanding the revenue that hunting can generate, this activity can also lead to several consequences in terms of disruption of age and social structure, animal dispersion and life expectancy [29,30]. Moreover, and as suggested in some studies [31,32], in wild populations, hunting may introduce an element of artificial selection. Thus, hunting activities can induce significant changes in demography and population genetics, such as social disruption or consequences on the gene pool of hunted populations [32,33]. The sustainability and conservation implications of this practice, which include effects on demography, genetic pool, behavior and general health, highlight the importance of management and regulation of hunting activities [5,27,33].

The main aim of this study was to investigate the impact of hunting activities on a wild population of red deer using measurements of glucocorticoids in plasma and hair and its metabolites in feces, all collected from hunted animals. We thus indirectly evaluated the stress levels of each hunted animal from a few months before the hunting event until its death. Given the wide effect that a stress response can have on the body, the physical and immunological condition of the animals was also evaluated.

2. Materials and Methods

2.1. Study Area and Red Deer Population

The study took place at Lousã Mountain (40°3′ N, 8°15′ W) and surrounding hunting areas, located in Central Portugal. The climate in this Mediterranean area is characterized by hot and dry summers and rainy winters [34]. With an altitude that ranges from 100 to 1205 m, this mountainous region is predominantly composed of plantations of coniferous and broadleaf trees, and large shrubland areas [26].

The red deer population in this region is the result of a reintroduction program that occurred between 1995 and 1999 with the release of 96 animals. Currently, this species occupies around 435 km^2 of the Lousã Mountain and surrounding areas, with an estimated density of 5.6 deer/km^2 during the rut season [35,36]. The calving season of this population occurs in May/June and rutting in mid-September to the end of October. With females being 37 ± 3% (mean ± SE) smaller than males, this species shows a very marked sexual dimorphism in body size. In addition, the sexual segregation outside the breeding season results in a matriarchal society wherein the females adopt a more philopatric behavior and males tend to disperse [35,37]. Since there are no natural predators, red deer are mainly preyed upon by feral dogs. This species is one of the most hunted big game species of the Lousã Mountain region, which includes 12 hunting areas, with red deer hunted in seven of them since the start of hunting in the region in 2006/2007 [26].

2.2. Data Collection

The study was performed during two hunting seasons (2013/2014 and 2014/2015), enabling the collection of samples from 80 red deer (38 adults: 26 females and 12 males; 30 sub-adults: 14 females and 16 males; and 12 young). The samples were collected during autumn (October and November) and winter (January and February) from red deer hunted in six "montarias" in three contiguous hunting areas: ZCM (Municipal Hunting Area) of Lousã (three hunting events—34 animals), ZCM of Vila Nova (one hunting event—7 animals) and ZCA (Associative Hunting Area) of Miranda do Corvo (two hunting events—39 animals) located in the Lousã Mountain area. No hunting events were conducted during the month of December. Post-mortem examination was made in situ, one to four hours after the animals were killed, and included the collection of blood, feces directly from the rectum, hair from the dorsal region and the metatarsus, and the recording of the sex and age class (based on animal size, body conformation and characteristics of antlers) of each animal. In addition, in November, fecal samples ($n = 11$) were also collected non-invasively during behavioral observations in the non-hunting area (i.e., area where hunting is not allowed) located in the central part of the Lousã Mountain, to use as the control.

Blood samples were taken directly from the heart into EDTA-tubes and centrifuged at 2000× g for 5 min. Plasma was collected and frozen at −20 °C in multiple aliquots. Feces, hair, and the metatarsus were frozen and stored at −20 °C for subsequent analyses.

2.3. Steroid Extraction and Quantification

Five ml of diethyl ether was added to each plasma sample (0.5 mL). After being shaken and centrifuged (2500× g, 15 min), the samples were frozen. Afterwards, the liquid component was transferred to a glass vial and dried under a stream of nitrogen (40 °C). This procedure was performed twice to increase the recovery of the extraction process. The combined and dried down extracts of each sample were dissolved in 0.5 mL of assay buffer and an aliquot analyzed by a cortisol enzyme immunoassay (EIA), as described in detail before [38]).

From each dried fecal sample 0.2 g were taken and mixed with 4 mL of methanol (100%) and 1 mL of water. The samples were shaken for 30 min and centrifuged at 2500× *g* for 15 min [39]. A group (with a 3α-11-one structure) of fecal cortisol metabolites (FCMs) was measured using an 11-oxoetiocholanonolone EIA. The utilized biotinylated label and the antibody (including its cross-reactivity) have been previously described in detail [40]. The assay has been successfully validated for use in red deer [41].

Hair samples were cut into fragments (<0.5 cm), washed with 5 mL 100% n-hexane to remove any lipids and potential external contamination, and air dried. Hair samples (0.05 g) were mixed with 5 mL of 100% methanol and incubated for 72 h for glucocorticoid extraction. After transferring the supernatant into a new glass vial, the organic solvent was evaporated at 40 °C using a stream of nitrogen. The extracts were dissolved in 0.5 mL of assay buffer and analyzed with a cortisol EIA [38,42].

2.4. Physical and Immunological Conditions

The physical condition of animals was assessed using bone marrow fat (BMF). The BMF was determined using the metatarsus, from where the bone marrow was extracted and weighed (± 0.0001 g). The bone marrow samples were then oven-dried at 60 °C and reweighed. BMF was determined as BMF = (the weight of oven-dried marrow/fresh marrow) × 100 [43].

To evaluate the immunological state of an individual, we used the blood collected in the field to make blood smears in the lab, followed by staining [44]. White blood cell counts and identification (100 cells per smears) were performed using a Nikon Eclipse Ni microscope (Nikon Instruments Europe B.V., Amsterdam, Netherlands). The examination and classification of blood cells were made based on morphological criteria and properties of staining, allowing the identification of five types of white blood cells, namely lymphocytes, neutrophils, eosinophils, monocytes and basophils, respectively [45].

2.5. Statistical Analysis

The correlation between the concentrations of GC in the different sampling tissues (i.e., plasma, feces and hair) was tested using Pearson's correlation. To analyse the physiological stress reactions, general linear models (LM) were used to test the effects of sex, age-class and month (independent variables) on the different GC levels (cortisol in plasma and hair, FCMs; dependent variables). Additionally, to evaluate the influence of the physical condition and the immunological status of the animals on GC levels, BMF and percentage of lymphocytes (the most abundant white blood cells (WBC)) were included in this analysis as independent variables. The concentrations of GC or its metabolites were transformed using a log transformation. BMF and WBC were logit transformed to achieve an approximation of a normal distribution and to reduce heterogeneity [46]. Since no significant interaction effects between the independent variables were found, only the main effects were used in the final models. The year had no significant effect on the concentrations of cortisol in plasma ($F_{(1,48)} = 1.989$; $p = 0.165$) or hair ($F_{(1,74)} = 0.022$; $p = 0.882$), nor on the concentrations of FCMs ($F_{(1,72)} = 2.691$; $p = 0.105$). Therefore, data of both years were pooled.

To evaluate the physical and immunological conditions of red deer, we tested the effects of sex, age-class and month (independent variables) on BMF and WBC (dependent variables) using general linear models (LM).

The results are expressed as mean ± SE and 95% confidence intervals (CI). All the statistical tests were considered significant when $p < 0.05$. The statistical analyses were performed using IBM.SPSS®, version 22 (IBM Corporation, New York, NY, USA).

3. Results

3.1. Physiological Stress Reactions

Concentrations of plasma and hair cortisol and its metabolites in feces were not correlated (Table 1). There was only a weak, but not significant correlation of cortisol in hair with fecal cortisol metabolites (FCMs).

Concentrations of plasma cortisol did not show differences between sex ($F_{(1,43)} = 0.361$; $p = 0.551$) or age classes ($F_{(2,43)} = 0.074$; $p = 0.929$). Comparison of plasma cortisol concentrations between months showed marginally significant differences ($F_{(3,43)} = 2.809$; $p = 0.051$) between the first and last months of the hunting season (Figure 1a). These results were not affected by BMF levels ($F_{(1,42)} = 0.029$; $p = 0.866$). However, the percentage of lymphocytes was associated with the concentration of plasma cortisol ($F_{(1,42)} = 5.947$; $p = 0.019$; $\beta = 2.553$). Once we controlled for the percentage of lymphocytes, the differences between months became more pronounced ($F_{(3,42)} = 4.356$; $p = 0.009$). FCM levels did not differ significantly in terms of sex ($F_{(1,67)} = 0.769$; $p = 0.384$), age class ($F_{(2,67)} = 1.589$; $p = 0.212$) or month ($F_{(3,67)} = 0.367$; $p = 0.777$; Figure 1b). Moreover, FCM levels were neither significantly influenced by the BMF levels ($F_{(1,65)} = 3.206$; $p = 0.078$) nor by the percentage of lymphocytes ($F_{(1,44)} = 1.151$; $p = 0.289$). Additionally, FCM values from hunted animals were higher (99.72 ± 11.33 ng/g) than those found in wild animals at control areas (51.17 ± 6.89 ng/g), for the same month (i.e., November). Regarding hair samples, levels of cortisol did not differ between sex ($F_{(1,69)} = 2.959$; $p = 0.090$) and age classes ($F_{(2,69)} = 1.570$; $p = 0.215$). Cortisol levels in hair differed between months ($F_{(3,69)} = 3.805$; $p = 0.014$), with the highest values recorded in February and the lowest in October (Figure 1c). Cortisol levels in hair were neither influenced by BMF ($F_{(1,68)} = 1.659$; $p = 0.202$) nor the percentage of lymphocytes ($F_{(1,46)} = 2.371$; $p = 0.130$), and the differences and patterns between months remained the same even after controlling for BMF levels and the percentage of lymphocytes.

Table 1. Pearson correlations between the concentrations of cortisol (metabolites) in plasma, feces and hair.

	Feces (*n* = 74)	**Hair (*n* = 76)**
Plasma (*n* = 50)	r = −0.152 *p* = 0.307	r = −0.085 *p* = 0.561
Feces (*n* = 74)	-	r = −0.217 *p* = 0.071

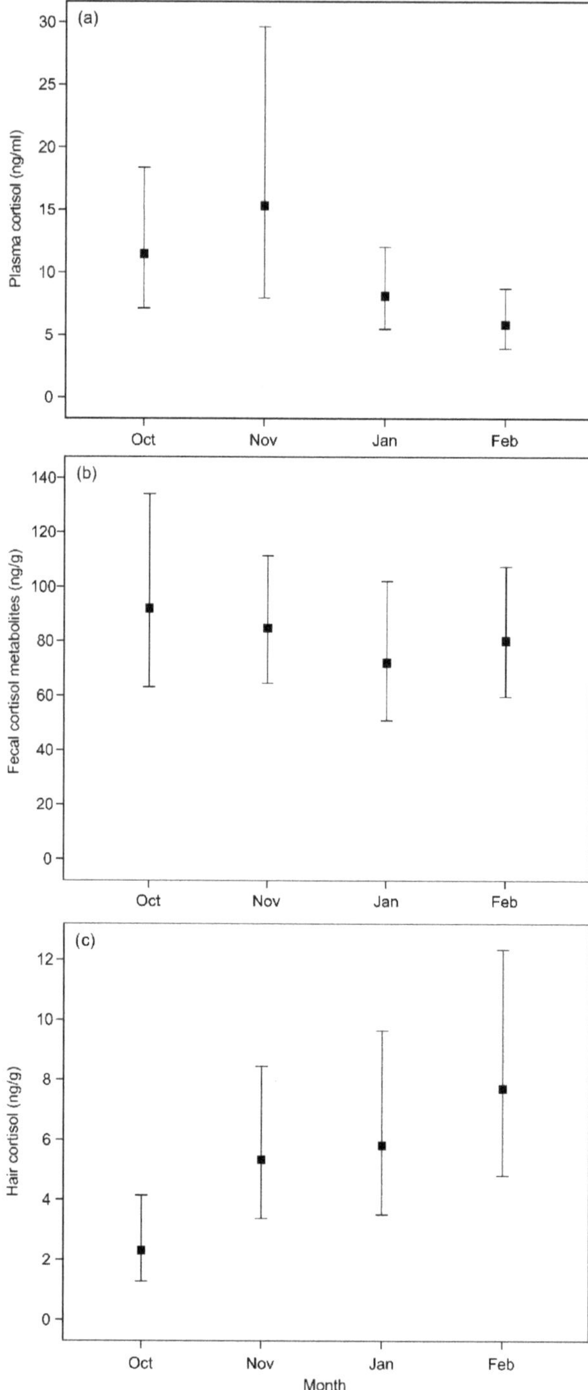

Figure 1. Patterns of cortisol (metabolite) levels in (**a**) plasma (**b**) feces and (**c**) hair during the hunting season (October to February). The values represent the estimated means from the general linear models and the bars represent the 95% confidence intervals.

3.2. Physical and Immunological Conditions

In terms of physical condition, measured by the BMF index, we found significant differences between the sexes ($F_{(1,72)} = 43.370$; $p < 0.001$), age classes ($F_{(2,72)} = 6.842$; $p = 0.002$) and months ($F_{(3.72)} = 3.288$; $p = 0.025$). Females had higher BMF values than males in all months (Figure 2). The differences obtained between age classes were mainly due to the poorer physical condition of calves (79.6; 95% CI [68.2, 87.6]) when compared with sub-adults (92.9; 95% CI [90.0, 94.9]) and adults (92.8; 95% CI [90.0, 94.8]). Regarding months, the animals had higher values of BMF in January (93.4; 95% CI [89.9, 95.8]) than in November (85.5; 95% CI [79.6, 90.0]). Lastly, white blood cells did not differ between sexes, age classes or across months ($p > 0.05$ for all models).

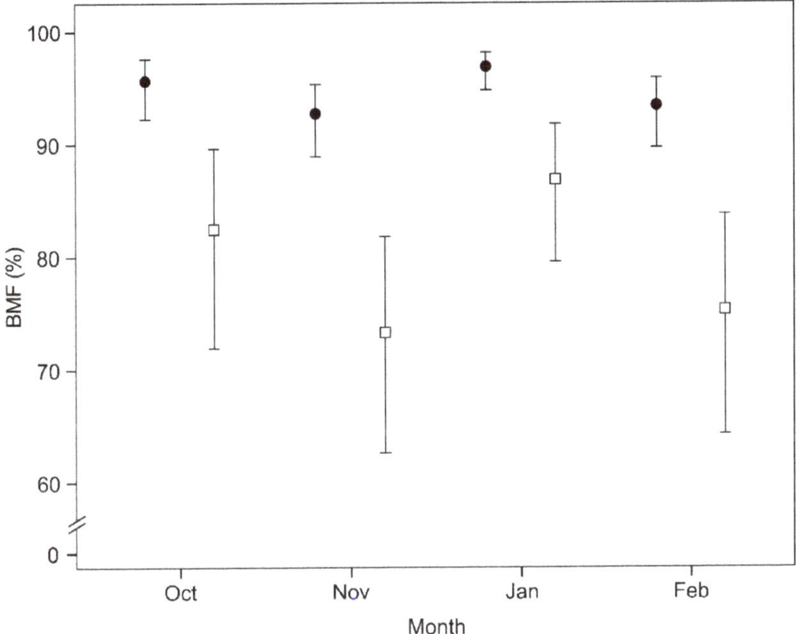

Figure 2. Patterns of bone marrow fat (BMF) levels for red deer females (black circles) and males (white squares) during the different months of the hunting season. The values represent the estimated means from the general linear models and the bars represent the 95% confidence intervals.

4. Discussion

Our results confirmed that concentrations of cortisol (or its metabolites) in plasma, feces and hair samples taken at the same time are not correlated. These results were expected since each sample matrix provides information about the endocrine state at different times. GC levels in plasma reflect an immediate physiological state [12], while feces provide information about the endocrine state a certain time before the sample collection [11] and hair is supposed to reflect the status of an accumulated period of some weeks to a few months [22]. Therefore, each biological sample matrix reflects distinct time-windows and thus complementary information can be gained.

The levels of cortisol and its metabolites did not differ between sex or age classes. These results were in agreement with those obtained by Huber et al. [17] who did not find differences in the concentration of FCMs between female and male red deer. However, other studies described sex-specific differences in glucocorticoid levels for some ruminant species [47,48] as well as age variations [49]. Regarding the influence of the reproductive state on stress levels, changes in cortisol levels were reported in red deer, with older females having higher cortisol levels in late gestation than non-pregnant females [50]. However, for reindeer (*Rangifer tarandus*) no differences were found in plasma cortisol concentrations between adult males, barren, and pregnant females [51]. Although the reproductive state of the females was not assessed in the present study, previous long-term data from the same study area has shown that during the sampling period more than 80% of females were usually pregnant (unpublished data), which means that probably more than 80% of the sampled females for this study were also pregnant. This fact, together with the absence of differences between males and females or age classes suggests that all animals were exposed to the same levels of stress during the sampling period. In fact, considering the type of hunting process used in our study area, which is not targeting any particular sex or age class, the obtained results are in agreement with our predictions. In fact, the absence of differences between males and females or age classes suggests that all animals were exposed to the same levels of stress during the sampling period.

We found a trend in plasma cortisol concentrations to decrease during the hunting season from November to February. A decrease of GC concentrations after regular and frequent occurrences of a stressor is often an indication of acclimation [2], as observed in some studies with Brahman cattle (*Bos taurus indicus*) and Magellanic penguins (*Spheniscus magellanicus*) [52,53]. However, besides the fact that we are dealing with a major stressor in our study, we need to consider the possible influence that this specific hunting method may have on our results. Since we have no information for how long deer were chased by the dogs before being killed, the cortisol values in plasma could reflect different chasing periods. Furthermore, the observed trend for plasma may also be influenced by other factors, such as food intake or/and any environmental disturbance [54–56], which make an interpretation more difficult. Although an influence of circadian rhythm in cortisol levels has been reported [55], the hunting events in the study area occurred within the same period of the day, decreasing the possible effect of daily circadian rhythm in our results.

Besides the trend obtained for plasma cortisol concentrations, the results also showed a positive correlation of plasma cortisol levels and the percentage of lymphocytes. We expected the opposite, namely a decrease of the percentage of lymphocytes with the increase of the stress levels, mainly due to the effect of stress as an immune suppressor [57]. Instead, our results were more in line with the immune-enhancing character of acute stress, promoting the passage of leukocytes from the blood to other parts of the body, while chronic stress induces immune suppression [57–59]. Although this interpretation requires caution given the existence of some unevaluated health-related factors, it points to the important relation between white blood cells and cortisol levels, and the need for more studies approaching this interaction.

We did not find differences in FCM concentrations across the sampled months. However, a seasonal pattern of GC levels in cervids is suggested by some authors who documented higher values in colder months than in warmer months of the year [17,60]. The minor influence of Mediterranean mild winters [34] where the occurrence of snow is uncommon and food availability is not significantly affected may have contributed to the lack of a seasonal pattern in our FCM levels during the hunting season. In fact, some studies in Mediterranean red deer [61] and roe deer (*Capreolus capreolus*) [62] suggested summer as the season with the most energetic constraints due to decreased food quality and quantity due to hydric stress. These results point, although weakly, given the small control sample size in terms of control group, to an influence of hunting in the stress levels. These results are in agreement with those reporting higher FCM levels in chamois (*Rupicapra rupicapra tatrica*) in areas with high disturbance than at low disturbance locations [16]. Hair cortisol concentrations differed significantly across months, with an increase from the beginning (October) to the end (February) of the hunting season. Cortisol levels recorded in February were significantly higher than the ones obtained in October. Taking into account that the molt from the summer to the winter coat is gradual, especially in adults where the development of new hair can occur before shedding the old one [63], our samples from October included new hair, which began to grow in September and October, and hair from the summer months which had not been shed yet. Furthermore, once hair follicle activity is reduced in

February, because the end of the winter season is getting close and the winter coat is fully grown [63], cortisol measured in hair that was sampled in this month should largely reflect the conditions from the previous three months [64]. Therefore, the increase in cortisol levels across months may be an indication of a period of prolonged stressful conditions induced by hunting activity, which is supported by the higher FCM levels in individuals from impact areas than from control sites found in our results. Similarly, Caslini et al. [65] found that hair cortisol levels in the same species (*Cervus elaphus*) were higher in greater density areas associated with more difficult environmental conditions and higher levels of anthropogenic disturbance (such as tourism). Our results are in agreement with those reported in the mentioned study which suggested that long-term HPA axis activity and allostatic load, as a consequence of higher densities, anthropogenic disturbances and/or environmental conditions in red deer populations, can be evaluated using cortisol hair levels as an index [65]. In addition, Bryan et al. [66] also documented higher hair cortisol levels in heavily hunted wolves (*Canis lupus*) than in wolves with lower hunting pressure. Hair cortisol seems to be a good indicator of long-term stress and has gained importance as a novel method to assess stressful conditions [19,22]. Hair can often be collected without capture and handling of the animals (i.e., hair traps). Collection of hair for hormone analyses may thus be a useful, non-invasive tool to monitor prolonged stressful conditions. Moreover, the fact that cortisol levels in hair may provide a long-term endocrine profile [67,68] can be extremely useful to study chronic stress and animal welfare [69]. On the other hand, taking into account our results regarding plasma cortisol levels and FCM concentrations, hair might not be suitable at capturing short-term stress levels.

Based on our results, and previous work, increased hair cortisol concentrations in our red deer population seem to be a consequence of hunting activities. However, considering that our study was focused on a wild population, there are other factors that may be contributing to the GC levels obtained in hair, like temperature, food availability or season [17,60,70]. As the energetic balance is a crucial factor in the ability of the animal to respond effectively to certain stressors [4], the body condition, a measure of the long-term energetic reserves [26,71], can have an influence on cortisol levels. Therefore, considering this bidirectional interaction, not only can stress affect body condition, energetic parameters could also be important in dealing with stress. Our results did not show any correlation between cortisol levels and BMF, which is a measure of physical condition, making the influence of season and/or food availability on cortisol levels unlikely. Moreover, the source of cortisol accumulated in hair is unclear, and some possible explanations have emerged. Keckeis et al. [72] reported a local production of GC in the hair follicles of guinea pigs, however, how this mechanism is modulated is still unknown. Recently, experimental evidence was provided in domestic sheep (*Ovis aries*) that mechanical irritation of the skin significantly increased hair cortisol concentrations [73]. Another study suggested the existence of a cutaneous HPA axis, able to synthesize and secrete cortisol, as well as negative feedback regulation by cortisol under corticotropin-releasing hormone (CRH) expression [74]. The uncertainty about the origin of cortisol in hair leads to an additional caution in the interpretation and analysis of GC levels in these types of samples [75,76]. To decrease the influence of confounding factors in our study, hair samples were taken from the dorsal region in all the individuals. However, further investigation would be very important to clarify whether cortisol concentrations are affected by the level of hair pigmentation as well as body area, hair type, or if there is any pattern along the hair shaft as suggested by some studies [18,20,21]. In addition, it is also relevant to emphasize the importance of the combined use of different indicators of stress to obtain more complete and precise information about the stress conditions of wild populations [10]. Plasma, feces and hair are complementary tools, which can provide information from different time-windows, allowing a better evaluation of the effects of human activities, like hunting, on the physiological stress response.

In terms of physical condition, our results showed that females were in better physical condition than males. This could be due to differential costs of reproduction for each sex, with males going through a phase of hypophagia and high activity levels during the rut [26,37]. Young individuals also had lower BMW indices than adults, which may be the result of a greater investment of these animals into growth [77]. Despite the observed differences in physical condition, and contrary to our predictions, the stress parameters we measured were not associated with physical condition. However, Cabezas et al. [78] reported lower values of body condition in animals with high GC levels in wild rabbits (*Oryctolagus cuniculus*). The absence of an association between stress levels and physical condition may indicate that the studied red deer population had enough fat reserves to cope with the stress induced by hunting activities.

5. Conclusions

The ability of plasma, feces and hair to provide multi-temporal information about physiological state proved to be very useful in the present study. Although the use of different biological samples increased the difficulty in the interpretation of our results, it allowed a broader panorama, which was more complete and reliable to understand given such a complex topic like stress reactions. We found evidence that repeated exposure of our red deer population to game hunting activity had an impact on stress levels, which can have important consequences for sustainability and conservation of this species. Specifically, stress can affect population dynamics, by changing foraging and breeding behavior, animal welfare, and, ultimately, the evolutionary processes, by changing individual fitness and selection [2,5,6]. Thus, exploring these topics is crucial to understanding the implications of hunting for the conservation of this species and to improve hunting management activities. Our study highlights the fundamental and broad role of stress in wildlife, emphasizing the need for more studies capable of clarifying how different biological matrices may be useful to evaluate the impacts of human pressure on wildlife, both in terms of stress level and stress processes.

Author Contributions: S.V., J.A. and A.A.d.S. conceived the experiments and performed the field work. S.V. and R.P. performed the laboratory analysis. J.A. and A.A.d.S. performed data analysis. S.V. and J.A. drafted the manuscript. S.V., A.A.d.S., R.P., K.E.R., J.P.S. and J.A. contributed to the writing, discussion and revision of the manuscript and have read and approved the final version of the manuscript. All authors have read and agreed to the published version of the manuscript.

Funding: This research was supported by POPH/FSE funds from the Portuguese Foundation for Science and Technology (FCT) through the fellowship of J. Alves (SFRH/BPD/123087/2016) and A. Alves da Silva (SFRH/BD/75018/2010). This study was funded by Instituto do Ambiente, Tecnologia e Vida da Faculdade de Ciências e Tecnologia da Universidade de Coimbra (IATV-UC) and supported by FCT—Fundação para a Ciência e a Tecnologia, I.P. within the project UID/BIA/04004/2019.

Acknowledgments: We thank the hunting areas managers of the ZCM of Lousã, ZCM of Vila Nova and ZCA of Miranda do Corvo for the collaboration during the collection of the biological samples. We also thank Filipe Rocha, Luis Pascoal da Silva and Fernanda Garcia for their help in the collection and preparation of the biological samples, Raquel Alves for helping with the classification of blood cells, and Marlon Millard for performing the EIA analyses.

Conflicts of Interest: The authors declare no conflicts of interest.

References

1. Möstl, E.; Palme, R. Hormones as indicators of stress. *Domest. Anim. Endocrinol.* **2002**, *23*, 67–74. [CrossRef]
2. Romero, L.M. Physiological stress in ecology: Lessons from biomedical research. *Trends Ecol. Evol.* **2004**, *19*, 249–255. [PubMed]
3. Martin, L.B. Stress and immunity in wild vertebrates: Timing is everything. *Gen. Comp. Endocrinol.* **2009**, *163*, 70–76. [CrossRef] [PubMed]
4. McEwen, B.S.; Wingfield, J.C. The concept of allostasis in biology and biomedicine. *Horm. Behav.* **2003**, *43*, 2–15. [CrossRef]
5. Busch, D.S.; Hayward, L.S. Stress in a conservation context: A discussion of glucocorticoid actions and how levels change with conservation-relevant variables. *Biol. Conserv.* **2009**, *142*, 2844–2853. [CrossRef]
6. Landys, M.M.; Ramenofsky, M.; Wingfield, J.C. Actions of glucocorticoids at a seasonal baseline as compared to stress-related levels in the regulation of periodic life processes. *Gen. Comp. Endocrinol.* **2006**, *148*, 132–149. [CrossRef] [PubMed]

7. Barja, I.; Silván, G.; Rosellini, S.; Piñeiro, A.; González-Gil, A.; Camacho, L.; Illera, J.C. Stress physiological responses to tourist pressure in a wild population of European pine marten. *J. Steroid Biochem. Mol. Biol.* **2007**, *104*, 136–142. [CrossRef] [PubMed]
8. Dantzer, B.; Fletcher, G.E.; Boonstra, R.; Sheriff, M.J. Measures of physiological stress: A transparent or opaque window into the status, management and conservation of species? *Conserv. Physiol.* **2014**, *2*, cou023. [CrossRef] [PubMed]
9. Palme, R. Measuring fecal steroids: Guidelines for practical application. *Ann. N. Y. Acad. Sci.* **2005**, *1046*, 75–80. [CrossRef] [PubMed]
10. Palme, R. Non-invasive measurement of glucocorticoids: Advances and problems. *Physiol. Behav.* **2019**, *199*, 229–243. [CrossRef] [PubMed]
11. Palme, R.; Rettenbacher, S.; Touma, C.; El-Bahr, S.M.; Möstl, E. Stress hormones in mammals and birds: Comparative aspects regarding metabolism, excretion, and noninvasive measurement in fecal samples. *Ann. N. Y. Acad. Sci.* **2005**, *1040*, 162–171. [CrossRef] [PubMed]
12. Sheriff, M.J.; Dantzer, B.; Delehanty, B.; Palme, R.; Boonstra, R. Measuring stress in wildlife: Techniques for quantifying glucocorticoids. *Oecologia* **2011**, *166*, 869–887. [CrossRef] [PubMed]
13. Boonstra, R.; Hubbs, A.H.; Lacey, E.A.; McColl, C.J. Seasonal changes in glucocorticoid and testosterone concentrations in free-living arctic ground squirrels from the boreal forest of the Yukon. *Can. J. Zool.* **2001**, *79*, 49–58. [CrossRef]
14. Kim, J.G.; Jung, H.S.; Kim, K.J.; Min, S.S.; Yoon, B.J. Basal blood corticosterone level is correlated with susceptibility to chronic restraint stress in mice. *Neurosci. Lett.* **2013**, *555*, 137–142. [CrossRef] [PubMed]
15. Formenti, N.; Viganó, R.; Fraquelli, C.; Trogu, T.; Bonfanti, M.; Lanfranchi, P.; Palme, R.; Ferrari, N. Increased hormonal stress response of Apennine chamois induced by interspecific interactions and anthropogenic disturbance. *Eur. J. Wildl. Res.* **2018**, *64*, 68. [CrossRef]
16. Zwijacz-Kozica, T.; Selva, N.; Barja, I.; Silván, G.; Martínez-Fernández, L.; Illera, J.C.; Jodlowski, M. Concentration of fecal cortisol metabolites in chamois in relation to tourist pressure in Tatra National Park (South Poland). *Acta Theriol.* **2013**, *58*, 215–222. [CrossRef]
17. Huber, S.; Palme, R.; Arnold, W. Effects of season, sex, and sample collection on concentrations of fecal cortisol metabolites in red deer (*Cervus elaphus*). *Gen. Comp. Endocrinol.* **2003**, *130*, 48–54. [CrossRef]
18. Macbeth, B.J.; Cattet, M.R.L.; Stenhouse, G.B.; Gibeau, M.L.; Janz, D.M. Hair cortisol concentration as a noninvasive measure of long-term stress in free-ranging grizzly bears (*Ursus arctos*): Considerations with implications for other wildlife. *Can. J. Zool.* **2010**, *88*, 935–949. [CrossRef]
19. Potratz, E.J.B.J.S.; Gallo, T.; Anchor, C.; Santymire, R.M. Effects of demography and urbanization on stress and body condition in urban white-tailed deer. *Urban. Ecosyst.* **2019**, *22*, 807–816. [CrossRef]
20. Dulude-de Broin, F.; Côté, S.D.; Whiteside, D.P.; Mastromonaco, G.F. Faecal metabolites and hair cortisol as biological markers of HPA-axis activity in the Rocky mountain goat. *Gen. Comp. Endocrinol.* **2019**, *199*, 229–243. [CrossRef] [PubMed]
21. Heimbürge, S.; Kanitz, E.; Otten, W. The use of hair cortisol for the assessment of stress in animals. *Gen. Comp. Endocrinol.* **2019**, *270*, 10–17. [CrossRef] [PubMed]
22. Russell, E.; Koren, G.; Rieder, M.; Van Uum, S. Hair cortisol as a biological marker of chronic stress: Current status, future directions and unanswered questions. *Psychoneuroendocrinology* **2012**, *37*, 589–601. [CrossRef] [PubMed]
23. Kalliokoski, O.; Jellestad, F.K.; Murison, R. A systematic review of studies utilizing hair glucocorticoids as a measure of stress suggests the marker is more appropriate for quantifying short-term stressors. *Sci. Rep.* **2019**, *9*, 11997. [CrossRef] [PubMed]
24. Bateson, P.; Bradshaw, E.L. Physiological effects of hunting red deer (*Cervus elaphus*). *Proc. Biol. Sci.* **1997**, *264*, 1707–1714. [CrossRef] [PubMed]
25. Gentsch, R.P.; Kjellander, P.; Röken, B.O. Cortisol response of wild ungulates to trauma situations: Hunting is not necessarily the worst stressor. *Eur. J. Wildl. Res.* **2018**, *64*, 11. [CrossRef]
26. Da Silva Alves, J.A. Ecological Assessment of the Red Deer Population in the Lous Lousã Mountain. Ph.D. Thesis, University of Aveiro, Aveiro, Portugal, 2013.
27. Torres-Porras, J.; Carranza, J.; Pérez-González, J. Selective culling of Iberian red deer stags (*Cervus elaphus hispanicus*) by selective montería in Spain European. *Eur. J. Wildl. Res.* **2009**, *55*, 117–123. [CrossRef]

28. Darimont, C.T.; Carlson, S.M.; Kinnison, M.T.; Paquet, P.C.; Reimchen, T.E.; Wilmers, C.C. Human predators outpace other agents of trait change in the wild. *Proc. Natl. Acad. Sci. USA* **2009**, *106*, 952–954. [CrossRef] [PubMed]
29. Allendorf, F.W.; England, P.R.; Luikart, G.; Ritchie, P.A.; Ryman, N. Genetic effects of harvest on wild animal populations. *Trends Ecol. Evol.* **2008**, *23*, 237. [CrossRef] [PubMed]
30. Milner, J.M.; Bonenfant, C.; Mysterud, A.; Gaillard, J.M.; Csányi, S.; Stenseth, N.C. Temporal and spatial development of red deer harvesting in Europe: Biological and cultural factors. *J. Appl. Ecol.* **2006**, *43*, 721–734. [CrossRef]
31. Coltman, D.W.; O'Donoghue, P.; Jorgenson, J.T.; Hogg, J.T.; Strobeck, C.; Festa-Bianchet, M. Undesirable evolutionary consequences of trophy hunting. *Nature* **2003**, *426*, 655–658. [CrossRef] [PubMed]
32. Martínez, J.G.; Carranza, J.; Fernández-García, J.L.; Sánchez-Prieto, C.B. Genetic variation of red deer populations under hunting exploitation in southwestern Spain. *J. Wildl. Manag.* **2002**, *66*, 1273–1282. [CrossRef]
33. Harris, R.B.; Wall, W.A.; Allendorf, F.W. Genetic consequences of hunting: What do we know and what should we do? *Wildl. Soc. Bull.* **2002**, *30*, 634–643.
34. Archibold, O.W. *Ecology of World Vegetation*; Chapman & Hall: London, UK, 1995.
35. Alves, J.; da Silva, A.A.; Soares, A.M.; Fonseca, C. Sexual segregation in red deer: Is social behaviour more important than habitat preferences? *Anim. Behav.* **2013**, *85*, 501–509. [CrossRef]
36. Alves, J.; da Silva, A.A.; Soares, A.M.; Fonseca, C. Spatial and temporal habitat use and selection by red deer: The use of direct and indirect methods. *Mamm. Biol.* **2014**, *79*, 338–348. [CrossRef]
37. Clutton-Brock, T.H.; Guinness, F.E.; Albon, S.D. *Red Deer: Behavior and Ecology of Two Sexes*; University of Chicago Press: Chicago, IL, USA, 1982.
38. Palme, R.; Möstl, E. Measurement of cortisol metabolites in faeces of sheep as a parameter of cortisol concentration in blood. *Z. Saugetierkd.* **1997**, *62*, 192–197.
39. Palme, R.; Touma, C.; Arias, N.; Dominchin, M.F.; Lepschy, M. Steroid extraction: Get the best out of faecal samples. *Wien. Tierarztl. Monatsschr. Vet. Med. Austria* **2013**, *100*, 238–246.
40. Möstl, E.; Maggs, J.L.; Schrötter, G.; Besenfelder, U.; Palme, R. Measurement of cortisol metabolites in faeces of ruminants. *Vet. Res. Commun.* **2002**, *26*, 127–139. [CrossRef]
41. Huber, S.; Palme, R.; Zenker, W.; Möstl, E. Non-invasive monitoring of the adrenocortical response in red deer. *J. Wildl. Manag.* **2003**, *67*, 258–266. [CrossRef]
42. Stubsjøen, S.M.; Bohlin, J.; Dahl, E.; Knappe-Poindecker, M.; Fjeldaas, T.; Lepschy, M.; Palme, R.; Langbein, J.; Ropstad, E. Assessment of chronic stress in sheep (part I): The use of cortisol and cortisone in hair as non-invasive biological markers. *Small Rumin. Res.* **2015**, *132*, 25–31. [CrossRef]
43. Neiland, K.A. Weight of dried marrow as indicator of fat in caribou femurs. *J. Wildl. Manag.* **1970**, *34*, 904–907. [CrossRef]
44. Mendes, J.; Gonçalves, A.C.; Alves, R.; Jorge, J.; Pires, A.; Ribeiro, A.; Sarmento-Ribeiro, A.B. L744, 832 and Everolimus Induce Cytotoxic and Cytostatic Effects in Non-Hodgkin Lymphoma. *Pathol. Oncol. Res.* **2015**, *22*, 301–309. [CrossRef] [PubMed]
45. Gautam, A.; Bhadauria, H. Classification of white blood cells based on morphological features. In Proceedings of the 2014 International Conference on Advances in Computing, Communications and Informatics (ICACCI), New Delhi, India, 24–27 September 2014; pp. 2363–2368.
46. Warton, D.I.; Hui, F.K. The arcsine is asinine: The analysis of proportions in ecology. *Ecology* **2011**, *92*, 3–10. [CrossRef] [PubMed]
47. He, L.; Wang, W.X.; Li, L.H.; Liu, B.Q.; Liu, G.; Liu, S.Q.; Qi, L.; Hu, D.F. Effects of crowding and sex on fecal cortisol levels of captive forest musk deer. *Biol. Res.* **2014**, *47*, 48. [CrossRef] [PubMed]
48. Munerato, M.S.; Marques, J.A.; Caulkett, N.A.; Tomás, W.; Zanetti, E.S.; Trovati, R.G.; Pereira, G.T.; Palme, R. Hormonal and behavioural stress responses to capture and radio-collar fitting in free-ranging pampas deer (*Ozotoceros bezoarticus*). *Anim. Welf.* **2015**, *24*, 437–446. [CrossRef]
49. Mooring, M.S.; Patton, M.L.; Lance, V.A.; Hall, B.M.; Schaad, E.W.; Fetter, G.A.; Fortin, S.S.; McPeak, K.M. Glucocorticoids of bison bulls in relation to social status. *Horm. Behav.* **2006**, *49*, 369–375. [CrossRef]
50. Pavitt, A.T.; Pemberton, J.M.; Kruuk, L.E.; Walling, C.A. Testosterone and cortisol concentrations vary with reproductive status in wild female red deer. *Ecol. Evol.* **2016**, *6*, 1163–1172. [CrossRef] [PubMed]

51. Bubenik, G.A.; Schams, D.; White, R.G.; Rowell, J.; Blake, J.; Bartos, L. Seasonal levels of metabolic hormones and substrates in male and female reindeer (*Rangifer tarandus*). *Comp. Biochem. Phys.* **1998**, *120*, 307–315. [CrossRef]
52. Franceschini, M.D.; Rubenstein, D.I.; Low, B.; Romero, L.M. Fecal glucocorticoid metabolite analysis as an indicator of stress during translocation and acclimation in an endangered large mammal, the Grevy's zebra. *Anim. Conserv.* **2008**, *11*, 263–269. [CrossRef]
53. Walker, B.G.; Boersma, P.D.; Wingfield, J.C. Habituation of adult Magellanic penguins to human visitation as expressed through behavior and corticosterone secretion. *Conserv. Biol.* **2006**, *20*, 146–154. [CrossRef]
54. Catanese, F.; Obelar, M.; Villalba, J.J.; Distel, R.A. The importance of diet choice on stress-related responses by lambs. *Appl. Anim. Behav. Sci.* **2013**, *148*, 37–45. [CrossRef]
55. Irvine, C.H.G.; Alexander, S.L. Factors affecting the circadian rhythm in plasma cortisol concentrations in the horse. *Domest. Anim. Endocrinol.* **1994**, *11*, 227–238. [CrossRef]
56. Wingfield, J.C. Ecological processes and the ecology of stress: The impacts of abiotic environmental factors. *Funct. Ecol.* **2013**, *27*, 37–44. [CrossRef]
57. Dhabhar, F.S.; McEwen, B.S. Acute stress enhances while chronic stress suppresses cell-mediated immunity in vivo: A potential role for leukocyte trafficking. *Brain Behav. Immun.* **1997**, *11*, 286–306. [CrossRef] [PubMed]
58. Dhabhar, F. Stress-induced changes in immune cell distribution and trafficking: Implications for immunoprotection versus immunopathology. In *Neural and Neuroendocrine Mechanisms in Host Defense and Autoimmunity*; Welsh, C.J., Meagher, M., Sternberg, E., Eds.; Springer: Berlin/Heidelberg, Germany, 2006; pp. 7–25.
59. Dhabhar, F.S. Stress-induced augmentation of immune function—The role of stress hormones, leukocyte trafficking, and cytokines. *Brain Behav. Immun.* **2002**, *16*, 785–798. [CrossRef]
60. Saltz, D.; White, G.C. Urinary cortisol and urea nitrogen responses to winter stress in mule deer. *J. Wildl. Manag.* **1991**, *55*, 1–16. [CrossRef]
61. Bugalho, M.N.; Milne, J.A. The composition of the diet of red deer (*Cervus elaphus*) in a Mediterranean environment: A case of summer nutritional constraint? *For. Ecol. Manag.* **2003**, *181*, 23–29. [CrossRef]
62. Minder, I. Local and seasonal variations of roe deer diet in relation to food resource availability in a Mediterranean environment. *Eur. J. Wildl. Res.* **2012**, *58*, 215–225. [CrossRef]
63. Ryder, M.L.; Kay, R. Structure of and seasonal change in the coat of red deer (*Cervus elaphus*). *J. Zool.* **1973**, *170*, 69–77. [CrossRef]
64. Loudon, A.; Milne, J.; Curlewis, J.; McNeilly, A. A comparison of the seasonal hormone changes and patterns of growth, voluntary food intake and reproduction in juvenile and adult red deer (*Cervus elaphus*) and Père David's deer (*Elaphurus davidianus*) hinds. *J. Endocrinol.* **1989**, *122*, 733–745. [CrossRef]
65. Caslini, C.; Comin, A.; Peric, T.; Prandi, A.; Pedrotti, L.; Mattiello, S. Use of hair cortisol analysis for comparing population status in wild red deer (*Cervus elaphus*) living in areas with different characteristics. *Eur J. Wildl Res* **2016**, *62*, 713–723. [CrossRef]
66. Bryan, H.M.; Jeg, S.; Koren, L.; Paquet, P.C.; Wynne-Edwards, K.E.; Musiani, M. Heavily hunted wolves have higher stress and reproductive steroids than wolves with lower hunting pressure. *Funct. Ecol.* **2015**, *29*, 347–356. [CrossRef]
67. Ventrella, D.; Elmi, A.; Barone, F.; Carnevali, G.; Govoni, N.; Bacci, M.L. Hair Testosterone and Cortisol Concentrations in Pre- and Post-Rut Roe Deer Bucks: Correlations with Blood Levels and Testicular Morphometric Parameters. *Animals* **2018**, *8*, 113. [CrossRef]
68. Ventrella, D.; Elmi, A.; Bertocchi, M.; Aniballi, C.; Parmeggiani, A.; Govoni, N.; Bacci, M.L. Progesterone and Cortisol Levels in Blood and Hair of Wild Pregnant Red Deer (*Cervus elaphus*) Hinds. *Animals* **2020**, *10*, 143. [CrossRef] [PubMed]
69. Macbeth, B.J.; Cattet, M.R.; Obbard, M.E.; Middel, K.; Janz, D.M. Evaluation of hair cortisol concentration as a biomarker of long.term stress in free-ranging polar bears. *Wildl. Soc. Bull.* **2012**, *36*, 747–758. [CrossRef]
70. Dantzer, B.; McAdam, A.G.; Palme, R.; Humphries, M.M.; Boutin, S.; Boonstra, R. How does diet affect fecal steroid hormone metabolite concentrations? An experimental examination in red squirrels. *Gen. Comp. Endocrinol.* **2011**, *174*, 124–131. [CrossRef] [PubMed]
71. Fuller, T.K.; Coy, P.L.; Peterson, W.J. Marrow fat relationships among leg bones of white-tailed deer. *Wildl. Soc. Bull.* **1986**, *14*, 73–75.

72. Keckeis, K.; Lepschy, M.; Schöpper, H.; Moser, L.; Troxler, J.; Palme, R. Hair cortisol: A parameter of chronic stress? Insights from a radiometabolism study in guinea pigs. *J. Comp. Physiol. B* **2012**, *182*, 985–996. [CrossRef] [PubMed]
73. Salaberger, T.; Millard, M.; El Makarem, S.E.; Möstl, E.; Grünberger, V.; Krametter-Frötscher, R.; Wittek, T.; Palme, R. Influence of external factors on hair cortisol concentrations. *Gen. Comp. Endocrinol.* **2016**, *233*, 73–78. [CrossRef]
74. Slominski, A.; Wortsman, J.; Tuckey, R.C.; Paus, R. Differential expression of HPA axis homolog in the skin. *Mol. Cell. Endocrinol.* **2007**, *265*, 143–149. [CrossRef]
75. Koren, L.; Bryan, H.; Matas, D.; Tinman, S.; Fahlman, Å.; Whiteside, D.; Smits, J.; Wynne-Edwards, K. Towards the validation of endogenous steroid testing in wildlife hair. *J. Appl. Ecol.* **2019**, *56*, 547–561. [CrossRef]
76. Jewgenow, K.; Azevedo, A.; Albrecht, M.; Kirschbaum, C.; Dehnhard, M. Hair cortisol analyses in different mammal species: Choosing the wrong assay may lead to erroneous results. *Conserv. Physiol.* **2020**, *8*. [CrossRef] [PubMed]
77. Servello, F.A.; Hellgren, E.C.; McWilliams, S.R. Techniques for wildlife nutritional ecology. In *Techniques for Wildlife Investigations and Management*; Braun, C., Ed.; The Wildlife Society: Bethesda, MD, USA, 2005; pp. 554–590.
78. Cabezas, S.; Blas, J.; Marchant, T.A.; Moreno, S. Physiological stress levels predict survival probabilities in wild rabbits. *Horm. Behav.* **2007**, *51*, 313–320. [CrossRef] [PubMed]

Article

Serum Protein Gel Agarose Electrophoresis in Captive Tigers

Daniela Proverbio [1,*], Roberta Perego [1,*], Luciana Baggiani [1], Giuliano Ravasio [1], Daniela Giambellini [2] and Eva Spada [1]

1 Department of Veterinary Medicine (DIMEVET), University of Milan, via dell'Università 6, 26900 Lodi, Italy; luciana.baggiani@unimi.it (L.B.); giuliano.ravasio@unimi.it (G.R.); eva.spada@unimi.it (E.S.)
2 Via Metteotti 9, Almenno San Salvatore, Bergam 24031, Italy; daniela.giambellini@gmail.com
* Correspondence: daniela.proverbio@unimi.it (D.P.); roberta.perego@unimi.it (R.P.); Tel.: +39-3482266335 (D.P.); +39-3388658384 (R.P.)

Received: 29 March 2020; Accepted: 17 April 2020; Published: 20 April 2020

Simple Summary: The tiger is the largest of the wild cats. There are fewer than 4000 wild tigers (*Panthera tigris*) worldwide and all subspecies of tigers are globally endangered. Due to pressures from poaching and retaliatory killings, this carnivore had lost an estimated 93% of its historic range. Given the critical situation of these wild felines, the health of each individual is of prime importance and laboratory blood testing, as well as evaluation of their physical condition, is important in their health assessment. Protein concentrations in the blood can be altered by malnutrition and dehydration as well as by disease. Serum electrophoresis allows the identification of the different protein fractions present in the blood and represents a useful tool in the diagnosis and monitoring of a number of diseases. Due to the nature of wild *Panthera tigris*, it is extremely difficult to obtain biological samples from free-living subjects, and therefore the values obtained from captive tigers provide very useful data. This study reports serum protein electrophoresis in 11 adult captive individuals. These results will be useful for the evaluation of physiological and pathological alterations in wild and captive tigers and populations.

Abstract: Given the endangered status of tigers (*Panthera tigris*), the health of each individual is important and any data on blood chemistry values can provide valuable information alongside the assessment of physical condition. The nature of tigers in the wild makes it is extremely difficult to obtain biological samples from free-living subjects, therefore the values obtained from captive tigers provide very useful data. Serum protein electrophoresis is a useful tool in the diagnosis and monitoring of a number of diseases. In this study, we evaluated agarose gel serum protein electrophoresis on samples from 11 healthy captive tigers. Serum electrophoresis on all 11 tiger samples successfully separated proteins into albumin, α_1, α_2, β_1, β_2 and γ globulin fractions as in other mammals. Electrophoretic patterns were comparable in all tigers. Mean± standard deviation or median and range values obtained for each protein fraction in healthy tigers were, respectively: 3.6 ± 0.2, 0.21 (0.2–0.23), 1.2 ± 0.2, 10.7 ± 0.2, 0.4 (0.3–0.6), 1.2 (1–1.8) gr/dL. The results of this preliminary study provide the first data on serum electrophoretic patterns in tigers and may be a useful diagnostic tool in the health assessment of this endangered species.

Keywords: *Panthera tigris tigris*; *Panthera tigris altaica*; siberian; tigers; bengal tigers; captive; biochemical parameter; serum protein electrophoresis

1. Introduction

The tiger is the largest of the wild cats [1]. This large carnivore has lost an estimated 93% of its historic range [2]. Furthermore, across their range, tigers face unrelenting pressures from poaching, retaliatory killings, and habitat loss. Fewer than 4000 wild tigers (*Panthera tigris*) are left in the world [3] and the International Union for Conservation of Nature (IUCN) has classified all subspecies of tigers (*Panthera tigris)* as globally endangered [2].

Approximately 10% of the world's tigers are found in the Russian Far East, where a single metapopulation represents the vast majority of Siberian, or Amur, tigers (*Panthera tigris altaica*). Today, there are fewer than 400 Amur tigers left in Russia and the eastern region of northeastern China [4–6]. The Bengal tiger (*Panthera tigris tigris*) is the most numerous of all tiger subspecies, with a wild population of more than 2500 [7].

In parallel, the world's largest population of tiger lives in captivity. Based on an estimate from a number of conservation organizations, there are approximately 10,000 tigers in captivity all over the world, with as many as 7000 tigers in the US in zoos, sanctuaries or privately owned [8]. Nevertheless, tigers in sanctuaries and zoological gardens worldwide represent a good source of animals for reintroduction into the wild which may play an increasingly important role in preventing the extinction of tigers through captive breeding programs [9]. Furthermore, such facilities are also central in educating the public about the critical status of the endangered tigers throughout the world [10].

Given the critical situation of these wild felines, maintaining the health of each individual is essential and any hematological and blood chemistry values that can provide information on the nutritional health status and physical condition are valuable. Furthermore, the detection of signs of disease in these animals is difficult and biological parameters are regarded as very useful complementary tools in the diagnosis and treatment of possible pathologies [11]. Total protein (TP) concentrations and protein fractions can be altered by several factors, such as dehydration, chronic malnutrition, malabsorption, maldigestion, protein-losing enteropathy, severe blood loss, chronic hepatic or renal disease, immunodeficiency, infectious or parasitic disease as well as metabolic or oncologic disorders [12,13]. Electrophoresis enables the separation of serum proteins into four/six fractions (albumin and α_1, α_2, β_1, β_2 and γ globulin fractions in order of decreasing anodal mobility), resulting in a typical electrophoretic pattern for the distribution of proteins [14]. Serum protein electrophoresis is considered one of the most reliable techniques for determining serum protein composition and, together with a basic hematologic and biochemical profile, are a useful tool in the diagnosis, prognosis, and monitoring of various diseases in both human and veterinary medicine [15]. In veterinary medicine, serum protein electrophoresis is mostly used for the investigation of hypoproteinemia and hyperproteinemia when screening for monoclonal or polyclonal gammopathies [16]. In domestic cats, electrophoretic pattern abnormalities are mainly associated with infectious/inflammatory diseases [17]. Wild cats can be affected by several disorders like infectious agents, including feline immunodeficiency virus (FIV), renal diseases, neoplastic and inflammatory changes [18]. Hypergammaglobulinemia related with myeloma has been reported in wild felid species like tiger [19,20] and lion [21]. Moreover, in captivity, tigers face many stressors and, even if the most effective way to objectively measure stress is by non-invasive measurement of stress hormone levels [9], evaluation of indirect markers of stress may also be useful [22] to ensure holistic wellness and health status [9].The acute phase proteins (APP), migrating in α_1 and α_2, globulin fractions, are a group of serum globulins, including ceruloplasmin, haptoglobin, a-2-macroglobulin, alpha1-acid-glycoprotein, that increase during acute inflammation, infection, surgical trauma or stress [22]. It has recently been suggested that phase APP may also be useful in the assessment of animal welfare [23].

The wild nature of *Panthera tigris* means it is extremely difficult to obtain biological samples from free-living subjects, making the data derived from captive individuals all the more useful. In addition, access to large numbers of free-living or captive individuals for testing is limited and researchers rely on data from a limited number of individuals [24,25].

To the knowledge of the authors, there is a paucity of literature regarding the serum electrophoretic pattern of *Panthera tigris.* Such data are valuable for the veterinary care of these animals. The objective of this study was to report values of total serum protein and electrophoresis fractions for healthy captive *Panthera tigris*, belonging to the subspecies *Panthera tigris tigris* (Bengala tiger) and *Panthera tigris altaica* (Siberian tiger). These results will be useful for the evaluation of physiological and pathological alterations in wild and captive tiger individuals and populations.

2. Materials and Methods

2.1. Animals and Sampling

Sera were collected from fifteen tigers, 6 Bengal tigers (*Panthera tigris tigris*), 7 tigers (*Panthera tigris)* and 2 Siberian tigers (*Panthera tigris Altaica*), including 7 neutered males and 8 neutered females, with ages ranging from 3.5 to 17 years. All tigers were housed, individually or in groups based on individual sociability, in various sized enclosures, rescue centers for exotic felids, zoological parks, or a circus located in northern Italy. Tigers were being immobilized for routine physical examination, ocular and dental examination, vaccination administration or minor surgical or diagnostic procedures. Each tiger received a clinical examination, and those with any visible abnormalities, with an inadequate body condition score, with signs of dehydration or with any signs of disease were excluded from the group of subjects considered clinically healthy.

2.2. Sample Preparation

As part of the health examination while under general anesthesia, blood samples were collected from the jugular, cephalic or saphenous vein of each animal. Blood was collected in plain tubes (Sistema BD Vacutainer®, Becton Dickinson Italia SPA, Italy) and serum was obtained by centrifugation at 10 min at 2500× *g*. Written owner consent for use of surplus blood samples, and use of data for scientific purposes, was obtained during consultations. The study design was approved by University of Milan Animal Welfare Bioethical Committee (Approval number OPBA 31/2019).

2.3. Total Serum Protein

Total serum protein concentration was measured by spectrophotometry using the colorimetric biuret method (Hagen Diagnostica S.R.L., Via Pratese 13 Firenze) on a Cobas Mira Classics Roche automated chemistry analyzer (Roche S.p.A., Mannheim, Germany).

2.4. Agarose Gel Electrophoresis

Sera samples were refrigerated at 4 °C and were analyzed within 8 hours of sampling. Protein fractions were analyzed using a semiautomated agarose gel electrophoresis (AGE) system (with HYDRAGEL Kit β_1-β_2 (SEBIA, Issy-les-Moulineaux, France). Serum was electrophoresed for 7 minutes at 33 volts hours and stained with diluted Amidoschwarz dye at pH 2 (4 g/L Amidoschwarz dye and 6.7% ethylene glycol). The AGE procedure was conducted according to the manufacturer's instructions, and commercial human serum was used as the control (normal control serum, Sebia, Evry, France). Using the computer software Phoresis for Windows 2000 or XP Pro (Sebia), the electrophoretic curve for each sample was displayed. Protein fractions were determined as the percentage optical absorbance, and the absolute concentration in g/dL was automatically calculated from the total serum protein concentration. Albumin to globulin (A/G) ratios were also calculated. The same operator analyzed all samples. To establish the inter-assay-accuracy of agarose gel electrophoresis on tiger serum, sera from two healthy tigers, one Bengal Tiger, 9A, F and one Siberian Tiger 1,1A, F, were tested 3 times on the same day, in the same laboratory and interpreted in duplicate by two operators. Coefficient of variability (CV) computed as SD/mean × 100 was calculated for each protein fraction.

2.5. Statistical Analysis

Data were tested for normality using the Shapiro–Wilk normality test. For normal distributions, means and standard deviations were calculated, and for non-normally distributed data, medians and ranges were calculated. Due to the scarcity of information on tiger electrophoretic patterns, values were compared to cheetah [13] and domestic cat [17] reference values. Due to the small sample size, the reference interval (RI) limits are directly estimated by the minimum and maximum values [26]. Statistical analysis was performed using MedCalc Statistical Software version 15.11.3 (MedCalc Software, Ostend 8400, Belgium).

3. Results

After clinical examination, only 11 out of 15 tigers were deemed clinically healthy. Therefore, serum samples from 11 tigers—five Bengal tigers (*Panthera tigris tigris*), five tigers (*Panthera tigris)* and one Siberian tiger (*Panthera tigris Altaica*), four neutered males and seven neutered females, with ages ranging from 5 to 16 years—were used to identify electrophoretic patterns of serum proteins in healthy tigers. Agarose gel electrophoresis carried out on all 11 tiger samples, successfully separated tiger serum proteins into albumin, α_1, α_2, β_1, β_2 and γ globulin fractions (Figure 1). All tigers had comparable electrophoretic patterns.

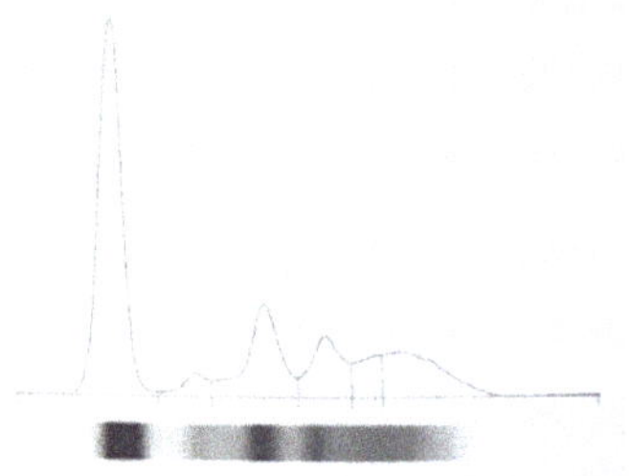

Figure 1. Electrophoretic curve of one healthy tiger. Agarose gel electrophoresis was able to separate serum proteins into six fractions: albumin, α1, α2, β1, β2 and γ globulin in order of decreasing anodal mobility.

All data, with the exception of α_1, β_2 and γ globulin fractions, were normally distributed. Gender-specific and age differences were not analyzed because the sample size was insufficient to allow statistical evaluation. Descriptive statistics of the protein serum electrophoresis fractions carried out in our healthy tiger population and reference values for cheetahs [13] and domestic cats [17] are given in Table 1.

The mean values of α_2 and β_1 globulin were 11.65% and 55.55% respectively, above the higher reference values indicated for domestic cats for the same globulin fractions [17]. Mean values of α_1, α_2 and γ globulin fractions were, respectively, 13.04% below the lower value and 74.24% and 18.18% above the higher value references indicated for cheetahs for the same globulin fractions [13]. The inter-assay accuracy of the agar gel electrophoresis in tiger serum was excellent as the same electrophoretic shape was recorded in all three repeated samples and for all protein fractions, with the exception of the α_1 globulin fraction in tiger number 2, the CVs were within the accepted ranges of within-subject biological variation for people (Table 2) [27].

Table 1. Total protein concentration and concentrations of albumin and α1, α2, β1, β2 and γ globulin fractions, obtained using agarose gel electrophoresis (AGE), in 11 healthy captive tigers. Mean, SD, minimum and maximum value. Reference value of electrophoretic fractions in cheetah and domestic cats from previous studies. ° data non-normally distributed.

Parameter	Mean +/− SD (range)	Minimum Value	Maximum Value	Reference Values Cheetah Depauw (2014)	Reference Values Domestic Cat Taylor (2010)
TP g/dL	7.4 ± 0.8	6.2	8.9		
Albumin g/dL	3.6 ± 0.2	3.3	3.9	3.2–4.8	2.9–4.67
α_1 globulin° g/dL	0.2 (0.2–0.23)	0.2	0.26	0.23–0.67	0.20–0.49
α_2 globulin g/dL	1.2 ± 0.2	0.8	1.5	0.13–0.66	0.29–1.03
$\beta_1$globulin g/dL	0.7 ± 0.2	0.44	1.2	0.4–0.8	0.15–0.45
β_2 globulin° g/dL	0.4 (0.3–0.6)	0.3	0.6	0.16–0.48	0.15–0.49
γ globulin° g/dL	1.2 (1–1.8)	0.8	2.3	0.29–1.1	0.43–2.14
A/G	0.92 ± 0.2	0.7	1.3	1.6	

Table 2. Serum protein electrophoresis in 2 healthy captive tigers serum samples tested 3 times on the same day, in the same laboratory and interpreted in duplicate by two operators. Coefficient of variability (CV) of total protein, albumin and α_1, α_2, β_1, β_2 and γ globulin calculated as SD/mean × 100.

Tiger	Total Protein g/dL	Albumin g/dL	$\alpha_1$globulin g/dL	$\alpha_2$globulin g/dL	$\beta_1$globulin g/dL	$\beta_2$globulin g/dL	γglobulin g/dL	A/G
Tiger 1	7	3.32	0.22	1.16	0.62	0.43	1.25	0.9
Tiger 1	7	3.19	0.23	1.23	0.62	0.41	1.33	0.84
Tiger 1	7	3.3	0.22	1.19	0.58	0.43	1.27	0.89
CV	0	2.14	2.27	2.94	3.83	2.61	3.2	4,8
Tiger 2	8.9	3.67	0.2	1.5	0.69	0.56	2.29	0.7
Tiger 2	8.9	3.89	0.15	1.41	0.66	0.5	2.3	0.78
Tiger 2	8.9	3.7	0.17	1.49	0.68	0.6	2.27	0.71
CV	0	3.2	14.4	3.4	1.7	9.1	0.6	5.4
CV_1		3.2	11.4	10.3	10.1	–	14.6	

CV_1: within-subject biologic variation for human samples (Westgard https://www.westgard.com/biodatabase1.htm).

4. Discussion

To the best of our knowledge, there have been no studies of the serum protein electrophoretic fractions in healthy tigers. The total serum protein electrophoretic pattern obtained with agarose gel electrophoresis separated the protein into six fractions, albumin, α_1, α_2, β_1, β_2 and γ globulins as in other mammals [14], resulting in a typical electrophoretic pattern for the distribution of proteins. Mean values of albumin and globulin fractions, with the exception of α_2 and β_1 globulins mean concentration, in our healthy tiger population fell within the reference values for domestic cats [17]. The average values of α_2 and β_1 globulins were above the higher reference values indicated for protein serum electrophoresis performed with AGE in domestic cats [17]. In mammals, the α_2 globulin fraction mainly consists of acute-phase proteins, such as α_1- acid glycoprotein, and often these proteins increase as a result of activation of the inflammatory response [28] to regulate different stages of inflammation [29]. Complement is one of the main proteins present in the β globulin fraction, corresponding to the sum of β_1 and β_2 globulin fractions [14]. Both α_2 and β_1 globulins fractions may be elevated if there is increased production of some acute phase proteins which migrate into these regions [14]. Recently, acute phase proteins have also been proposed as useful stress biomarkers. In humans, cows and experimental animals, psychological and physical stress elevates plasma interleukin-6 and APP levels [22,30]. Acute phase proteins are synthesized predominantly in the liver, in response to secretion of pro-inflammatory cytokines. In response to stress signals, the hypothalamic–pituitary–adrenal (HPA) axis may trigger cytokine production resulting in an increase in hepatic APP synthesis and release into the bloodstream [22]. Although lacking specificity, the detection of an increase in α_2 globulin could help in monitoring the stress status of tigers in captivity.

Although the sample size analyzed was limited, the data obtained in this study could suggest that healthy tigers may have a higher concentration of acute phase proteins than domestic cats, or that the tiger population studied could have been in an inflammatory state. The interpretation of serum protein electrophoretic patterns depends on the variations among different groups of animals. Moreover, acute phase proteins are a variable group of serum proteins and concentrations vary widely between different animal species [31]. Domestic cats are in the same family and share a similar physiology to tigers and could be an acceptable alternative for comparison of normal values for many biochemical parameters; however, they are a different subfamily and extrapolation of all results is dangerous. Depauw et al. (2014) [13] reported results of captive cheetah (*Acinonyx jubatus*) serum protein electrophoretic fractions by capillary electrophoresis (CE). Although the AGE and CE are different techniques for protein fraction separation, the shape of the electrophoretogram of cheetah serum was comparable to that found in our healthy tiger sample [13].

The percentage of variation observed between protein fractions of two healthy tiger serum samples repeatedly submitted to agarose gel electrophoresis to evaluate inter-assay-accuracy were within the accepted ranges of within-subject biological variation for people [27]. Only the α_1 globulin fraction exceeded the acceptable value of 20.83%. This result could be due to the low concentration of α_1 globulins in the serum. In fact, the accuracy of analysis is usually better for protein fractions found in higher serum concentrations because low concentrations are more susceptible to small changes [32]. Cushing et al. (2019) [19] described a cases series of myeloma associated with hypergammaglobulinemia in five adult tigers. Diagnosis of myeloma is based on a variety of clinical signs often associated with monoclonal gammopathy found in serum. It is interesting to note that in this case series, the serum protein electrophoresis was done in the absence of reference values for the serum protein pattern typical of this species. This underlines the importance of acquiring a database of reference values even for the rarest wild carnivores.

There were a number of limitations with this study. Firstly, although each tiger was clinically examined, and screened for visible alterations and low body condition score, the history was sometimes incomplete or unavailable, which may have compromised the accurate categorization of animals according to health status and disease type. In addition, samples were from animals in different types of housing, so the diversity of habitat and diet could have affected the results. [25]. Furthermore, due to the small number of subjects, we are not able to define reference ranges. In fact, following the reference interval guidelines of the American Society for Veterinary Clinical Pathology, reference ranges should not be calculated when the sample size is <20 subjects. In these cases, mean or median and minimum and maximum values should be provided [33]. For the same reason, we did not evaluate the influence of gender or age on serum protein electrophoretic patterns.

In veterinary medicine, serum protein electrophoresis is recognized as a useful tool in the diagnosis, prognosis and monitoring of a number of diseases [15,17]. Alterations of blood biochemical values are interpreted by comparison with the same value in healthy subjects. This study provides useful data on the serum protein electrophoretic values in healthy captive tigers that increases our understanding of this endangered species. Furthermore, the paucity of reports of variations in serum protein electrophoretic patterns in free-ranging or captive tigers makes this preliminary study a useful aid for the evaluation of physiological and pathological alterations in both wild and captive tiger populations.

5. Conclusions

Serum protein electrophoresis is a useful tool in the diagnosis and monitoring of a number of diseases. This study presents serum protein electrophoresis in a sample of healthy captive tigers. Due to the nature of wild *Panthera tigris*, it is extremely difficult to obtain biological samples from free-living subjects, and therefore the values obtained from captive tigers provide very useful data. Results indicate that agarose gel electrophoresis separates the total serum protein into six fractions, albumin,

α_1, α_2, β_1, β_2 and γ globulins in tigers as in other mammals, resulting in a typical electrophoretic pattern for the distribution of proteins. Mean values of albumin and globulin fractions, with the exception of α_2 and β_1 globulins mean concentration, in our healthy tiger population fell within the reference values indicated for protein serum electrophoresis performed with AGE in domestic cats. These preliminary results provide the first data on serum electrophoretic pattern in healthy tigers and may offer a platform for further research into serum proteins as a useful diagnostic tool in the health assessment of this endangered species.

Author Contributions: Conceptualization, D.P., E.S., R.P.; methodology, L.B. and D.G.; formal analysis, D.P..; investigation, L.B.; resources and samples G.R.; data curation, D.P.; E.S. and R.P.; writing—original draft preparation, D.P.; writing—review and editing, D.P., E.S., R.P., L.B.; project administration, D.P. All authors have read and agreed to the published version of the manuscript.

Funding: This research received no external funding

Conflicts of Interest: The authors declare no conflict of interest

References

1. Sajjad, S.; Farooq, U.; Malik, H.; Anwar, M.; Ahmad, I. Comparative hematological variables of Bengal tigers (Panthera tigris tigris) kept in lahore Zoo and Lahore Wildlife Park, Pakistan. *Turk. J. Vet. Anim. Sci.* **2012**, *36*, 346–351.
2. Peng, Z.; Ning, Y.; Liu, D.; Sun, Y.; Wang, L.; Zhai, Q.; Hou, Z.; Chai, H. Ascarid infection in wild Amur tigers (Panthera tigris altaica) in China. *BMC Vet. Res.* **2020**, *16*, 1–7. [CrossRef] [PubMed]
3. Walston, J.; Robinson, J.G.; Bennett, E.L.; Breitenmoser, U.; da Fonseca, G.A.B.; Goodrich, J.; Gumal, M.; Hunter, L.; Johnson, A.; Ullas Karanth, K.; et al. Bringing the tiger back from the brink-the six percent solution. *PLoS Biol.* **2010**, *8*, e1000485. [CrossRef] [PubMed]
4. Ning, Y.; Kostyria, A.V.; Ma, J.; Chayka, M.I.; Guskov, V.Y.; Qi, J.; Sheremetyeva, I.N.; Wang, M.; Jiang, G. Dispersal of Amur tiger from spatial distribution and genetics within the eastern Changbai mountain of China. *Ecol. Evol.* **2019**, *9*, 2415–2424. [CrossRef] [PubMed]
5. Carroll, C.; Miquelle, D.G. Spatial viability analysis of Amur tiger Panthera tigris altaica in the Russian Far East: The role of protected areas and landscape matrix in population persistence. *J. Appl. Ecol.* **2006**, *43*, 1056–1068. [CrossRef]
6. Kerley, L.L.; Mukhacheva, A.S.; Matyukhina, D.S.; Salmanova, E.; Salkina, G.P.; Miquelle, D.G. A comparison of food habits and prey preference of Amur tiger (Panthera tigris altaica) at three sites in the Russian Far East. *Integr. Zool.* **2015**, *10*, 354–364. [CrossRef]
7. Seidensticker, J. Saving wild tigers: A case study in biodiversity loss and challenges to be met for recovery beyond 2010. *Integr. Zool.* **2010**, *5*, 285–299. [CrossRef]
8. Hartigan, R. Are Wildlife Sanctuaries Good for Animals? Available online: Nationalgeographic.com/news/2011/110320 animal-sanctuary-wildlife-exotic-tiger-zoo/ (accessed on 15 April 2020).
9. Narayan, E.J.; Parnell, T.; Clark, G.; Martin-Vegue, P.; Mucci, A.; Hero, J.M. Faecal cortisol metabolites in Bengal (Panthera tigris tigris) and Sumatran tigers (Panthera tigris sumatrae). *Gen. Comp. Endocrinol.* **2013**, *194*, 318–325. [CrossRef]
10. Kelly, P.; Stack, D.; Harley, J. A review of the proposed reintroduction program for the far eastern leopard (Panthera pardus orientalis) and the role of conservation organizations, veterinarians, and zoos. *Top. Companion Anim. Med.* **2013**, *28*, 163–166. [CrossRef]
11. Helena, M.; Akao, M.; Mieko, R.; Mirandola, S.; Ito, F.H.; Itikawa, P.H.; Pessoa, R.B. Hematologic Parameters of Captive Lions. *Acta Sci. Vet.* **2015**, *55*, 1–6.
12. Proverbio, D.; de Giorgi, G.B.; Pepa, A.D.; Baggiani, L.; Spada, E.; Perego, R.; Comazzi, C.; Belloli, A. Preliminary evaluation of total protein concentration and electrophoretic protein fractions in fresh and frozen serum from wild Horned Vipers (Vipera ammodytes ammodytes). *Vet. Clin. Pathol.* **2012**, *41*, 582–586. [CrossRef] [PubMed]

13. Depauw, S.; Delanghe, J.; Whitehouse-Tedd, K.; Kjelgaard-Hansen, M.; Christensen, M.; Hesta, M.; Tugirimana, P.; Budd, J.; Dermauw, V.; Janssens, G.P.J. Serum Protein Capillary Electrophoresis and Measurement of Acute Phase Proteins in a Captive Cheetah (Acinonyx Jubatus) Population. *J. Zoo Wildl. Med.* **2014**, *45*, 497–506. [CrossRef] [PubMed]
14. Tothova, C.; Nagy, O.; Kovac, G. Serum proteins and their diagnostic utility in veterinary medicine: A review. *Vet. Med.* **2016**, *61*, 475–496. [CrossRef]
15. Giordano, A.; Paltrinieri, S. Interpretation of capillary zone electrophoresis compared with cellulose acetate and agarose gel electrophoresis: Reference intervals and diagnostic efficiency in dogs and cats. *Vet. Clin. Pathol.* **2010**, *39*, 464–473. [CrossRef] [PubMed]
16. Gerou-Ferriani, M.; Mcbrearty, A.R.; Burchmore, R.J.; Jayawardena, K.G.I.; Eckersall, P.D.; Morris, J.S. Agarose gel serum protein electrophoresis in cats with and without lymphoma and preliminary results of tandem mass fingerprinting analysis. *Vet. Clin. Pathol.* **2011**, *40*, 159–173. [CrossRef]
17. Taylor, S.S.; Tappin, S.W.; Dodkin, S.J.; Papasouliotis, K.; Casamian-Sorrosal, D.; Tasker, S. Serum protein electrophoresis in 155 cats. *J. Feline Med. Surg.* **2010**, *12*, 643–653. [CrossRef]
18. Junginger, J.; Hansmann, F.; Herder, V.; Lehmbecker, A.; Peters, M.; Beyerbach, M.; Wohlsein, P.; Baumgärtner, W. Pathology in captive wild felids at German zoological gardens. *PLoS ONE* **2015**, *10*, e0130573. [CrossRef]
19. Cushing, A.C.; Sc, B.V.; Cert, A.V.P.Z.M.; Dipl, A.C.Z.M.; Ramsay, E.C.; Newman, S.J.; Dipl, A.C.V.P.; Hespel, A.M. Hypergammaglobulinemia and Myeloma in Five Tigers (Panthera Tigris): Clinicopathological Findings. *J. Zoo Wildl. Med.* **2019**, *50*, 219.
20. Lee, A.M.; Guppy, N.; Bainbridge, J.; Jahns, H. Multiple myeloma in an amur tiger (Panthera tigris altaica). *Open Vet. J.* **2017**, *7*, 300–305. [CrossRef]
21. Tordiffe, A.S.W.; Cassel, N.; Lane, E.P.; Reyers, F. Multiple myeloma in a captive lion (panthera leo). *J. South Afr. Vet. Assoc.* **2013**, *84*, 1–5. [CrossRef]
22. Murata, H.; Shimada, N.; Yoshioka, M. Current research on acute phase proteins in veterinary diagnosis: An overview. *Vet. J.* **2004**, *168*, 28–40. [CrossRef]
23. Piñeiro, M.; Piñeiro, C.; Carpintero, R.; Morales, J.; Campbell, F.M.; Eckersall, P.D.; Toussaint, M.J.M.; Lampreave, F. Characterisation of the pig acute phase protein response to road transport. *Vet. J.* **2007**, *173*, 669–674. [CrossRef] [PubMed]
24. Farooq, U.; Sajjad, S.; Anwar, M.; Khan, B.N. Serum Chemistry Variables of Bengal Tigers (Panthera tigris tigris) Kept in Various Forms of Captivity Study area and experimental animals: The present Standard capture and sampling protocol: Standard capture protocol was used and observed at both sit. *Pak. Vet. J.* **2011**, *8318*, 283–285.
25. Shrivatav, A.B.; Singh, K.P.; Mittal, S.K.; Malik, P.K. Haematological and biochemical studies in tigers (Panthera tigris tigris). *Eur. J. Wildl. Res.* **2012**, *58*, 365–367. [CrossRef]
26. Wiesel, I.; Biol, D.; Zimmerman, D.M.; Kirk, W.; Dipl, A.C.Z.M.; Le Boedec, K.; Deem, S.; Griot-Wenk, M.E.; Giger, U.; Junbo Zhang, S.Y. Reference interval estimation of small sample sizes: A methodologic comparison using a computer-simulation study. *Vet. Clin. Pathol.* **2018**, *49*, 335–346.
27. Spada, E.; Proverbio, D.; Baggiani, L.; Canzi, I.; Perego, R. Hematological reference values for stray colony cats of northern Italy: Hematological references intervals for stray cats. *Comp. Clin. Path.* **2016**, *25*. [CrossRef]
28. O'Connell, T.X.; Horita, T.J.; Kasravi, B. Understanding and interpreting serum protein electrophoresis. *Am. Fam. Physician* **2005**, *71*, 105–112.
29. Petersen, H.H.; Nielsen, J.P.; Heegaard, P.M.H. Application of acute phase protein measurements in veterinary clinical chemistry. *Vet. Res.* **2004**, *35*, 163–187. [CrossRef]
30. Kim, M.H.; Yang, J.Y.; Upadhaya, S.D.; Lee, H.J.; Yun, C.H.; Ha, J.K. The stress of weaning influences serum levels of acute-phase proteins, iron-binding proteins, inflammatory cytokines, cortisol, and leukocyte subsets in Holstein calves. *J. Vet. Sci.* **2011**, *12*, 151–158. [CrossRef]
31. Eckersall, P.D.; Bell, R. Acute phase proteins: Biomarkers of infection and inflammation in veterinary medicine. *Vet. J.* **2010**, *185*, 23–27. [CrossRef]

32. Rosenthal, K.L.; Johnston, M.S.; Shofer, F.S. Assessment of the reliability of plasma electrophoresis in birds. *Am. J. Vet. Res.* **2005**, *66*, 375–378. [CrossRef] [PubMed]
33. Friedrichs, K.R.; Harr, K.E.; Freeman, K.P.; Szladovits, B.; Walton, R.M.; Barnhart, K.F.; Blanco-Chavez, J. ASVCP reference interval guidelines: Determination of de novo reference intervals in veterinary species and other related topics. *Vet. Clin. Pathol.* **2012**, *41*, 441–453. [CrossRef] [PubMed]

Article

Circadian Rhythm of Salivary Immunoglobulin A and Associations with Cortisol as A Stress Biomarker in Captive Asian Elephants (*Elephas maximus*)

Tithipong Plangsangmas [1,2], Janine L. Brown [3], Chatchote Thitaram [2,4], Ayona Silva-Fletcher [5], Katie L. Edwards [3,6], Veerasak Punyapornwithaya [7], Patcharapa Towiboon [2] and Chaleamchat Somgird [2,4,*]

1 Master's Degree Program in Veterinary Science, Faculty of Veterinary Medicine, Chiang Mai University, Chiang Mai 50100, Thailand; tithipong_p@cmu.ac.th
2 Center of Elephant and Wildlife Research, Faculty of Veterinary Medicine, Chiang Mai University, Chiang Mai 50100, Thailand; chatchote.thitaram@cmu.ac.th (C.T.); towiboon@gmail.com (P.T.)
3 Center for Species Survival, Smithsonian Conservation Biology Institute, Front Royal, VA 22630, USA; BrownJan@si.edu (J.L.B.); k.edwards@chesterzoo.org (K.L.E.)
4 Department of Companion Animal and Wildlife Clinics, Faculty of Veterinary Medicine, Chiang Mai University, Mae Hia, Chiang Mai 50100, Thailand
5 Department of Clinical Sciences and Services, The Royal Veterinary College, Hawkshead Lane, Hertfordshire AL9 7TA, UK; ASilvaFletcher@rvc.ac.uk
6 North of England Zoological Society, Chester Zoo, Upton-by-Chester, CH2 1LH, UK
7 Veterinary Public Health and Food Safety Centre for Asia Pacific (VPHCAP), Faculty of Veterinary Medicine, Chiang Mai University, Chiang Mai 50100, Thailand; veerasak.p@cmu.ac.th
* Correspondence: chaleamchat.s@cmu.ac.th; Tel.: +66-53948-015

Received: 18 December 2019; Accepted: 13 January 2020; Published: 17 January 2020

Simple Summary: Salivary immunoglobulin A (sIgA) and cortisol concentrations were measured in Asian elephants to determine circadian rhythm effects and the relationship between both biomarkers. Saliva samples were collected every 4 h from 06:00 to 22:00 h for 3 consecutive days (n = 15 samples/elephant). We used enzyme immunoassays for quantification of sIgA and cortisol concentrations. Both sIgA and cortisol followed a circadian rhythm, although the patterns differed. For both, the highest concentrations were in the early morning hours when elephants began the work day; however, sIgA concentrations were more variable during the day. There was no correlation between the two indices because the pattern of sIgA was quartic, while that of cortisol was linear. We provide basic knowledge for further studies using sIgA as a welfare biomarker.

Abstract: Salivary immunoglobulin A (sIgA) has been proposed as a potential indicator of welfare for various species, including Asian elephants, and may be related to adrenal cortisol responses. This study aimed to distinguish circadian rhythm effects on sIgA in male and female Asian elephants and compare patterns to those of salivary cortisol, information that could potentially have welfare implications. Subjects were captive elephants at an elephant camp in Chiang Mai province, Thailand (n = 5 males, 5 females). Salivette® kits were used to collect saliva from each elephant every 4 h from 06:00 to 22:00 h for 3 consecutive days (n = 15 samples/elephant). Enzyme immunoassays were used to quantify concentrations of IgA and cortisol in unextracted saliva. Circadian rhythm patterns were determined using a generalized least-squares method. Both sIgA and cortisol followed a circadian rhythm, although the patterns differed. sIgA displayed a daily quartic trend, whereas cortisol concentrations demonstrated a decreasing linear trend in concentrations throughout the day. There was no clear relationship between patterns of sIgA and salivary cortisol, implying that mechanisms of control and secretion differ. Results demonstrate for the first time that circadian rhythms affect sIgA, and concentrations follow a daily quartic pattern in Asian elephants, so standardizing time of collection is necessary.

Keywords: Asian elephant; saliva; immunoglobulin A; circadian rhythm; glucocorticoids; welfare

1. Introduction

The Asian elephant (*Elephas maximus*) is the official national animal of Thailand, classified as endangered by the International Union for Conservation of Nature (IUCN 2010), and listed in Appendix 1 of the Convention on International Trade in Endangered Species of Wild Fauna and Flora (CITES). Since the logging ban in 1986, thousands of elephants and their mahouts were left without work and took to the streets to beg for food. About a decade later, a new industry emerged for Thai elephants—tourism. In 2017, there were 2673 elephants working in 223 tourism venues throughout the country (National Elephant Institute, Lampang, Thailand). While a few camps offer observation only, most utilize elephants in a variety of scheduled activities, like riding with and without saddles, entertainment shows, and tourist feeding and bathing. Around 900 elephants reside in more than 80 venues in Chiang Mai province. Variation in elephant demographics, work activities, elephant care, and mahout management is evident among the camps of northern Thailand, as recently reviewed by Bansiddhi et al. [1], all of which can affect how individuals cope with the tourist environment.

Assessment of animal welfare relies on measures of physiological function (e.g., health, reproduction, stress) and/or behavior, applied at individual or population levels. The most commonly used biomarkers of stress and, by extension welfare, are glucocorticoids (GC) [2]. In response to an acute stimuli, activation of the hypothalamic-pituitary-adrenal (HPA) axis causes the release of cortisol from the adrenal cortex [3], which then feeds back to inhibit further release to restore homeostasis [4]. In humans and animal models, cortisol is typically measured in blood serum or plasma; however, the potential for inducing stress due to handling and blood collection [5] is a concern for most wildlife species, especially for repeated sampling. Thus, noninvasive approaches that quantify GC metabolites in urine or feces have been developed to assess acute and chronic stress responses in many species [6,7]. Another method—saliva collection—is less invasive than blood and, with a lag time of only 20–30 min, provides almost real time results [7]. Although studies have shown the value of GC for monitoring stress and welfare, including in elephants [8,9], today it is recognized that additional indicators that incorporate multiple physiological systems offer more ways to assess both negative and positive welfare states [10].

Recently, salivary immunoglobulin A (sIgA) has been promoted as a potential biomarker of positive affects [11,12]. Immunoglobulin A is found in many secretory fluids, including saliva and breast milk, and in nasal, gastrointestinal, bronchial, and urogenital secretions [13]. In general, sIgA responds quickly to acute events (positive or negative), increasing or decreasing depending on the stressor. Positive mood inductions related to movies, music, and self-referent statements have been shown to increase sIgA [14], as well as relaxation and massages [13]. However, sIgA concentrations also alter in response to negative effects. For example, depletion of sIgA occurs in humans taking academic exams [15,16], while elevations have been related to mental arithmetic tasks, and reported daily hassles and work demands [13,17]. Studies in mammalian species have linked sIgA to stress as well. In dogs, noise stressors and defense training caused a decrease in sIgA [18,19], while puppies displayed increased sIgA after behavioral testing [19,20]. In pigs, sIgA increased due to restraint stress [21] and isolation [22]. However, before using sIgA as a welfare biomarker, baseline levels must be established, as well as any endogenous patterns. Two studies have measured immunoglobulin A in Asian elephants [23,24], but neither examined specific temporal patterns.

The circadian rhythm is a roughly 24 h cycle in physiological processes, which generally are endogenously generated, although they can be modulated by recurring external cues, such as sunlight, temperature, and sleep-wake and activity cycles [25]. The suprachiasmatic nucleus (SCN) in the hypothalamus serves as the master pacemaker that sets the timing of rhythms by regulating neuronal activity, body temperature, and hormonal signals [26]. Circadian patterns are important to consider when interpreting biological results, to distinguish between basal rhythms and extrinsic effects. It is generally accepted that most mammals exhibit circadian patterns in GC secretion, with concentrations being highest in the morning and lowest at around midnight [16,22,27,28]. Studies in Asian elephants [29,30] have confirmed this pattern in urine [29] and saliva [30,31] samples. IgA also has been shown to have a diurnal pattern. In humans, concentrations decline throughout the day from a morning peak at 08:00–09:00 h. By contrast, in pigs [21] and dogs [18], sIgA concentrations are lowest in the morning (09:00 h), increase during the day (11:00–15:00 h), and then decline at night (17:00 h). There are no reports of circadian rhythms in sIgA in Asian elephants, nor its relationship with the well-studied stress hormone, cortisol. Thus, the goals of this study were to assess temporal patterns of sIgA throughout the day as a potential novel biomarker aiding in the assessment of welfare in elephants, and compare the patterns to those of cortisol. We hypothesized that sIgA in Asian elephants follows a circadian rhythm that is correlated with salivary cortisol.

2. Materials and Methods

2.1. Animals and Sample Collection

All animal procedures were approved by the Institutional Animal Care and Use Committee, Faculty of Veterinary Medicine, Chiang Mai University, Chiang Mai, Thailand (license number; S2/2561).

Saliva samples were collected from five male and five female Asian elephants aged 34.5 ± 4.7 years (range, 11–54 years) and weighed 3216 ± 306.29 kg (range, 2568–3702 kg) by applying the heart girth equation [32] from an elephant facility in Chiang Mai, Thailand. Samples were collected every 4 h between 06:00 to 22:00 h for 3 consecutive days (n = 15 samples/elephant). Bull elephants were restrained with long chains (30 m) during the day and short chains (5 m) at night. Female elephants were kept unrestrained in a fenced area (1600 m^2) allowing social interactions during the day (09:00–16:00 h) and on short chains (5 m) at night. Elephants at night were chained inside an open shed with other elephants in close proximity, but with no physical contact. Females participated in tourist feeding and bathing routines twice a day (from 09:00–12:00 and 13:00–16:00 h), while bulls did not interact with tourists. Saliva collection did not interfere with the normal routine of the elephants during the day. At night, lights were turned off at around 22:00 h to allow elephants to rest, so samples were not collected between 22:00 and 06:00 h. Saliva was collected using Salivette® kits (Sarstedt, AG and Co, Numbrecht, Germany) by swiping the absorbent piece inside the buccal area for 30–60 s, which took less than 5 min to complete. Samples were kept in 4 °C coolers for less than 8 h until centrifuged at 1500× g for 5 min at 15 °C. Two swabs were collected and the saliva pooled, resulting in an average volume of 500 µL (100–1500 µL) per sample. Saliva was stored at −30 °C until analysis. Samples were analyzed within 3 months as suggested by Ng et al. [33].

2.2. Enzyme Immunoassays

2.2.1. Immunoglobulin A

Immunoglobulin A was quantified in Asian elephant saliva by enzyme immunoassay (EIA) as described by Edwards et al. [24] with some modifications. A polyclonal rabbit anti-human IgA antibody (A0262, Dako, Glostrup, Denmark) was diluted to a working concentration of 1 mg/L in phosphate buffered saline (0.01 M phosphate buffer, 0.15 M NaCl, pH 7.2) (PBS) and 100 µL added per well to a 96-well microtiter plate (Nunc-Immuno maxisorp, Thermo Fisher Scientific, Roskilde, Denmark). After incubation overnight at 4 °C, plates were aspirated and washed three times with phosphate buffered saline with tween (PBS-T). Standards (0.39–100 µg/L; I2636, Sigma Aldrich, St. Louis, MO, USA) and saliva samples diluted 1:100 in PBS-T were added in duplicate. Following incubation at room temperature (RT) for 2 h on a plate shaker set to 150 rpm, plates were aspirated and washed three times with PBS-T. A polyclonal rabbit anti-human IgA antibody conjugated to horseradish peroxidase (HRP; P0216, Dako, Glostrup, Denmark) was diluted 1:10,000 in PBS-T and 100 µL added per well before incubation at room temperature (RT) for 1 h on a plate shaker set to 150 rpm. After a final wash step, 100 μL of 3,3′,5,5′-tetramethylbenzidine (TMB) was added per well and incubated in the dark for 10 min at RT. Finally, the reaction was stopped with 50 μL stop solution (1N HCl) and the absorbance measured at 450 nm using a microplate reader (TECAN Sunrise, Salzburg, Austria). Assay sensitivity was 3.37 ng/mL. The EIA was validated for elephant saliva by demonstrating parallelism between serial dilutions of saliva and the IgA standards ($y = 7.8042x + 0.2779$, $R^2 = 0.986$) and significant recovery of IgA added to low concentration saliva before analysis ($y = 0.935x + 0.485$, $R^2 = 0.997$). Samples were analyzed in duplicate; inter-assay coefficient of variation (CV) was 10.49% (n = 3), and the intra-assay CV was 2.36%.

2.2.2. Cortisol

Concentrations of salivary cortisol were determined using a double-antibody EIA with a polyclonal rabbit anti-cortisol antibody (R4866). Second antibody-coated plates were prepared by adding 150 μL of anti-rabbit IgG (0.01 mg/mL) to each well of a 96-well microtiter plate (Nunc-Immuno maxisorp, Thermo Fisher Scientific, Roskilde, Denmark), and incubated at RT for 15–24 h. The wells were then emptied and blotted dry, followed by adding 250 μL blocking solution (100 mM phosphate, 150 mM sodium chloride, 1% Tween 20, 0.09% sodium azide, 10% sucrose, pH 7.5) and incubating for 15–24 h at RT. After incubation, all wells were emptied, blotted, and dried at RT in a desiccating cabinet (Sanpla Dry Keeper, Sanplatec Corp., Auto A-3, Japan) with loose desiccant in the bottom. After drying (humidity < 20%), plates were heat-sealed in a foil bag with a 1g desiccant packet and stored at 4 °C until use. Neat samples (50 μL) or cortisol standards (50 μL) were added to appropriate wells. Cortisol-horseradish peroxidase (HRP) (25 μL) was immediately added to each well, followed by 25 μL anti-cortisol antibody (except non-specific binding wells) and incubated at RT for 1 h on a plate shaker set to 150 rpm. Plates were then washed four times with wash buffer (1:20 dilution, 20× Wash Buffer Part No. X007; Arbor Assays, Ann Arbor, MI, USA) and 100 μL of TMB dihydrochloride dissolved in phosphate-citrate buffer with sodium perborate (Sigma Aldrich, St. Louis, MO, USA) was added, followed by incubation for 10 min at RT without shaking. The reaction was stopped with 50 μL stop solution (1N HCl) and absorbance was measured at 450 nm by a microplate reader (TECAN Sunrise, Salzburg, Austria). Assay sensitivity based on 90% binding was 0.084 ng/mL. The cortisol EIA was validated for elephant saliva by demonstrating parallelism between serial dilutions of saliva and the cortisol standards ($y = -10.946x + 99.705$, $R^2 = 0.996$) and significant recovery of cortisol added to low concentration saliva before analysis ($y = 0.7935x + 0.0743$, $R^2 = 0.9987$). Samples were analyzed in duplicate; inter-assay and intra assay CVs were 5.48% (n = 4) and 1.46%, respectively.

2.3. *Statistical Analysis*

All analyses were performed using R statistical software version 3.5.1 [34]. Descriptive data were reported as mean ± standard error of the mean (SEM) for both sIgA and cortisol concentrations in each time period, and as overall concentrations for each sex. A generalized least-squares method (GLS) was used to compare differences in sIgA and cortisol concentrations over time. The GLS model was constructed by using nonlinear mixed-effects (nlme) package 3.1-137 [35]. We constructed the model using time period and day of sample collection as the main effects. Individual elephant was defined as a random effect. For GLS modelling, the Akaike information criterion (AIC) was determined from models with different covariance structures, including compound symmetry, autoregressive process of order 1 (AR1), and general correlation matrix with no structure. The compound symmetry structure had the lowest AIC value for the sIgA comparison model and AR1 had the lowest AIC value for cortisol, indicating the best fitted models. Therefore, the structure of the covariance pattern for GLS models for sIgA and cortisol were defined as the compound symmetry and AR1, respectively. Significant differences in mean sIgA and cortisol concentrations between different time periods were analyzed by Tukey's post-hoc tests followed by examining linear, quadratic, and quartic trend effects over the 24 h cycle using the linear regression model and the locally weighted least squares regression (loess) method. Residuals from the fitted models were tested for normality and homogeneity of variance assumption by plotting standardized residuals versus quantiles of standard normal (QQ normality graph) and plotting standardized residuals versus fitted values, respectively. The plot indicated no violation for both assumptions, thus transformation of the sIgA and cortisol concentration data was not necessary. The scatter plots of sIgA and cortisol values were created using ggplot2 package 3.1.1 [36]. The repeated measures correlation (rmcorr) package 0.3.0 [37] was used to determine the correlation between sIgA and cortisol accounted for inter-individual differences in baseline concentrations. For all statistical tests, the significance level was set at $\alpha = 0.05$.

3. Results

Average sIgA and cortisol concentrations for the three collection periods are summarized in Table 1. The highest average sIgA and cortisol concentrations were observed in samples collected at 06:00 h, while the lowest values occurred between 18:00 and 22:00 h, respectively. Individual concentrations of every time point is provided in Supplementary Materials Table S1.

Table 1. Daily and overall mean ± standard error of the mean (SEM) concentrations and ranges of salivary immunoglobulin A (sIgA) and cortisol in 10 Asian elephants (n = 5 male, 5 female).

Parameter (ng/mL)	Day	Time (hours) 06:00	10:00	14:00	18:00	22:00	Min-Max (ng/mL)
sIgA	1	60.61 ± 3.99	53.40 ± 5.63	43.69 ± 4.00	55.52 ± 6.62	55.04 ± 6.38	22.80–83.70
	2	85.13 ± 11.4	61.62 ± 6.88	71.33 ± 7.27	51.89 ± 6.82	79.13 ± 7.75	19.87–150.18
	3	70.36 ± 8.68	42.85 ± 6.03	58.05 ± 5.4	39.17 ± 6.17	49.17 ± 4.92	13.71–96.19
Overall		72.09 ± 5.07 [a]	52.96 ± 3.69 [b]	56.67 ± 3.57 [a,b]	48.64 ± 3.76 [b]	61.75 ± 4.29 [a,b]	13.71–150.18
Cortisol	1	0.82 ± 0.09	0.76 ± 0.10	0.86 ± 0.15	0.71 ± 0.31	0.55 ± 0.15	0.12–3.17
	2	0.87 ± 0.17	0.56 ± 0.14	0.52 ± 0.05	0.34 ± 0.10	0.48 ± 0.18	0.08–2.01
	3	0.63 ± 0.06	0.49 ± 0.14	0.41 ± 0.10	0.28 ± 0.05	0.23 ± 0.06	0.10–1.26
Overall		0.79 ± 0.07 [a]	0.61 ± 0.07 [a,b]	0.59 ± 0.08 [a,b]	0.46 ± 0.12 [a,b]	0.45 ± 0.09 [b]	0.08–3.17

[a,b] Means within rows with different superscripts are significantly different for each biomarker ($p < 0.05$).

Average sIgA and cortisol concentrations by sex are summarized in Table 2, with no differences observed at each time point (sIgA: $p = 0.57$, Cortisol: $p = 0.73$).

Table 2. Comparison of overall mean (± SEM) concentrations of salivary immunoglobulin A (sIgA) and cortisol between sexes (n = 5 males, 5 females) throughout three, 24 h periods.

Parameter (ng/mL)	Time (hours) 06:00	10:00	14:00	18:00	22:00	Min-Max (ng/mL)
sIgA						
Male	65.88 ± 8.55	52.77 ± 8.34	59.83 ± 7.11	46.06 ± 6.91	60.93 ± 6.45	13.71–125.12
Female	78.74 ± 8.84	53.15 ± 3.58	52.72 ± 4.77	51.62 ± 6.15	62.7 ± 8.7	14.88–150.18
Cortisol						
Male	0.83 ± 0.13	0.69 ± 0.12	0.5 ± 0.08	0.37 ± 0.07	0.42 ± 0.14	0.08–2.01
Female	0.73 ± 0.10	0.52 ± 0.11	0.71 ± 0.15	0.61 ± 0.29	0.44 ± 0.12	0.1–3.17

From the GLS model, the effect of time was significant for sIgA ($p = 0.0001$), but only approached significance for cortisol ($p = 0.06$). There was an effect of collection day (sIgA; $p < 0.0001$, cortisol; $p = 0.0235$) for both biomarkers. Even though no effect of time was found for cortisol in the model, post hoc comparisons using the Tukey's honestly significant difference (HSD) test indicated that mean sIgA concentration at 06:00 h was higher than that at 10:00 h ($p = 0.002$) and 18:00 h ($p = 0.0001$). By contrast, mean cortisol concentration at 06:00 h was only higher than that at 22:00 h ($p = 0.0373$). All data were used to construct a trend line using loess regression analysis. Mean and standard deviation (SD) are presented in Figures 1 and 2 as well as the trend line for IgA and cortisol concentrations. Quartic (p = 0.04) trend effects for sIgA (Figure 1), and linear (p = 0.002) trend effects for cortisol (Figure 2) were evident throughout the 24 h cycle.

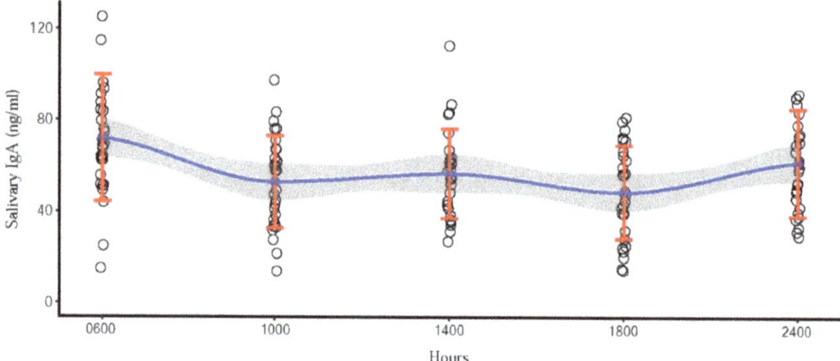

Figure 1. Scatter plot indicating quartic circadian pattern of sIgA concentrations in 10 Asian elephants. Blue dots represent the mean concentration of each time point. Red error bars represent the standard deviation. Blue line represents the trend line, and the shaded area is a 95% confidence interval of the mean.

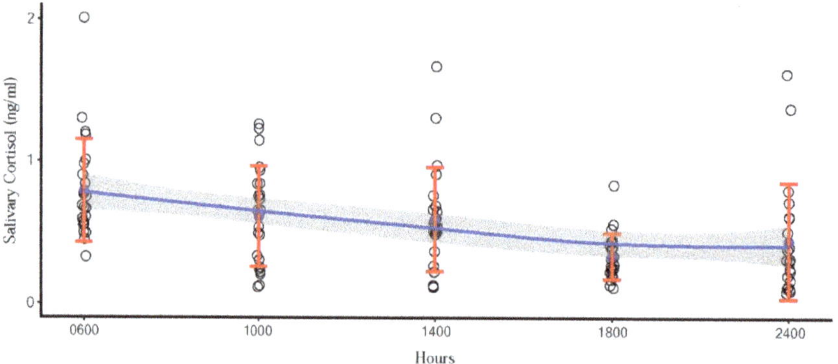

Figure 2. Scatter plot indicating linear circadian pattern of cortisol concentrations in 10 Asian elephants. Blue dots represent the mean concentration of each time point. Red error bars represent the standard deviation. Blue line represents the trend line, and the shaded area is a 95% confidence interval of the mean.

sIgA and cortisol patterns of representative individuals depicting quartic and linear effects of the mean are shown in Figures 3 and 4, respectively. However, there was considerable variability and not all elephants followed clear patterns, as indicated in Figure 5. sIgA and cortisol patterns of all individuals is provided in Supplementary Materials Figures S1 and S2. There were missing samples at certain time points because of insufficient volume of saliva for analysis due to a combination of agitated elephants and human error in the collection process (Figures 4 and 5). There was only a weak, non-significant positive correlation (r = 0.099; $p > 0.05$) between sIgA and cortisol.

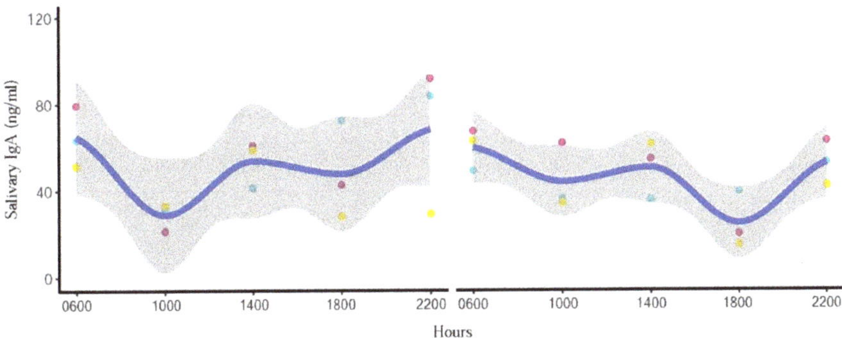

Figure 3. Representative individual profiles showing quartic circadian IgA trends in one male (**left**) and one female (**right**) elephant. Blue line represents the trend line, and the shaded area is a 95% confidence interval. Day 1 = red points, Day 2 = blue points, and Day 3 = yellow points.

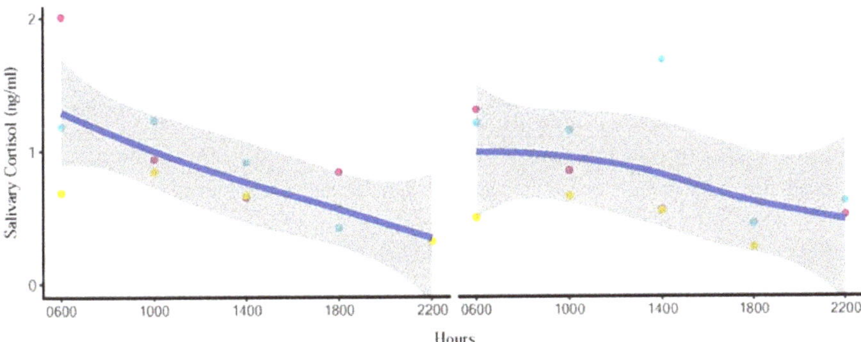

Figure 4. Representative individual profiles showing linear circadian cortisol trends in one male (**left**) and one female (**right**) elephant. Blue line represents the trend line, and the shaded area is a 95% confidence interval. Day 1 = red points, Day 2 = blue points, and Day 3 = yellow points.

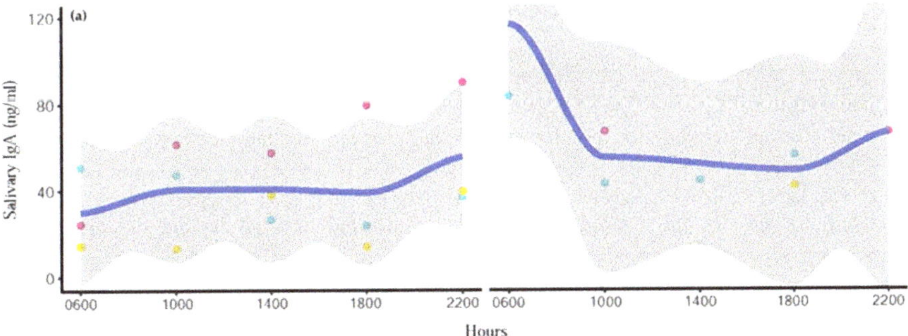

Figure 5. *Cont.*

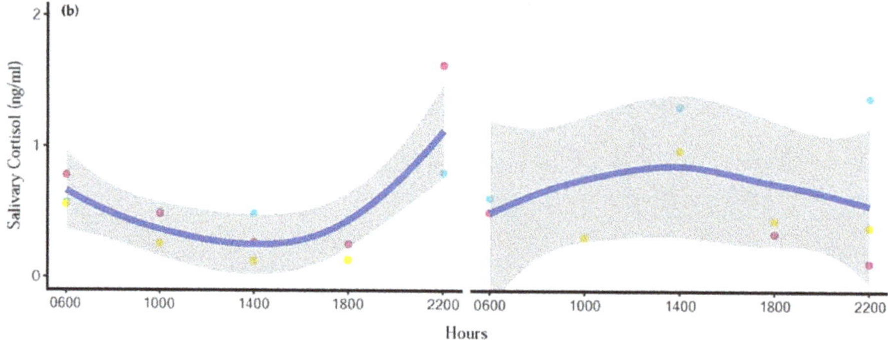

Figure 5. Representative profiles of diurnal (**a**) IgA and (**b**) cortisol trends in one male (left) and one female (right) elephant that did not follow the typical or expected pattern. Blue line represents the trend line, and the shaded area is a 95% confidence interval of the mean. Day 1 = red points, Day 2 = blue points, and Day 3 = yellow points.

4. Discussion

This is the first study to measure both salivary IgA and salivary cortisol in Asian elephants throughout the day (16-h period), and we found that the trend line from regression analysis followed a diurnal pattern; however, the two were not significantly correlated because cortisol followed a linear pattern, whereas that of sIgA was quartic. The highest concentrations of cortisol were observed at 06:00 h, with the lowest at 22:00 h. By contrast, sIgA concentrations were elevated at 06:00 and 22:00 h, with nadirs at 10:00 and 18:00 h.

Circadian rhythms of IgA are known to vary among species [12]. In humans, sIgA concentrations peak in the morning and decrease throughout the day [38,39]. One study that measured sIgA concentrations over a full 24-h period revealed a gradual increase in sIgA starting at midnight with peak concentrations occurring at 08:00 h the following day [27]. Shirakawa et al. [27] recorded patterns of sIgA in humans coinciding with the sleep-wake time of the subjects, which were between 24:00 and 07:00 h. Our study showed similar peaks in concentrations for sIgA at 06:00 h, which was about an hour after wake time for the elephants. Although the slight increase in sIgA at 14:00 h in our study was not significantly different from 06:00 or 22:00 concentrations, it also was not different from 10:00 and 18:00 h, which agrees with the quartic pattern observed in humans. For elephants in this study, the sleep hours are between 23:00 and 05:00 h. The sIgA increase in samples collected at 22:00 h in elephants corresponded to the time when elephants began their standing sleep period, and agrees with changes associated with the sleep-wake cycle in humans [38]. By contrast, in pigs [21] and dogs [18], peak concentrations are observed during the afternoon. Authors speculate that species-related behavior and differences in daily routines could be the cause of contrasting patterns between species. Like humans, the daily activities of captive elephants are generally fixed, with tourist activities in the morning and afternoon, and a break in the middle of the day. Elephants often sleep in a standing position during these rest periods, which might explain the slight increase in sIgA at 14:00 h. Dogs also show peak concentrations in conjunction with intermittent sleep during afternoon hours [18]. However, the bulls in this study did not interact directly with tourists, yet still showed a quartic pattern, perhaps because although contact was limited, they were still aware of tourist presence. Bull elephants did have a daily routine with mahouts that bathed and fed them at regular intervals, which also could have driven a circadian pattern. Because we were unable to measure a full 24-h cycle, further studies of sIgA concentrations in elephants during sleep would be beneficial to determine the complete cycle and confirm its resemblance to that of humans as compared to other species.

IgA has been measured in multiple sample types (blood, saliva, urine, and feces) across time, including samples from Asian elephants [24]. Samples in the study of Edwards et al. [24] were collected only once a day, with no apparent attention to time, so circadian patterns were not determined. However, concentrations were highly variable, especially for feces and saliva, with serum having the lowest variability. Concentrations in urine were low, with many being undetectable, suggesting it may not be the best sample type to assess this biomarker. In addition to within animal variability, Edwards et al. [24] also found considerable between animal differences, similar to our study. Overall individual mean concentrations ranged from ~7 to 30 ng/mL saliva for five elephants in Edwards et al. [24], and ~41–70 ng/mL for 10 elephants in the present study. Understanding mechanisms driving this significant intra- and inter-animal variability is key to understanding the utility of IgA as a potential health or welfare biomarker.

sIgA concentrations differed somewhat between the present study and that of Edwards et al. [24] in that our overall mean sIgA was more than double in concentration. Possible causes could be related to minor modifications in the assay protocol, which included using lower antibody (1 mg/L versus 10 mg/L) and HRP (1:10,000 versus 1:2500 dilutions) concentrations, although the standard curve range was the same between studies. Climate and daylight hour differences between regions (Washington DC versus Chiang Mai) could have had an effect [40]. For example, Park and Tokura [41] found that brighter light conditions during the day resulted in higher concentrations of sIgA during nocturnal sleep in humans. From Mishra et al. [42], people that travelled from India to Antarctica exhibited increases in sIgA concentrations, which could reflect differences in either climate or day length. No visual differences were evident over six months of longitudinal sIgA data between February and August [24], whereas the Thailand study was conducted in August, so any influence of seasonality may be minimal. Other factors associated with sIgA secretion are age [19,43–47], sex [45], and health status [48,49]. Concentrations of sIgA did not differ between sexes or were related to age in this study, whereas Edwards et al. [24] found the highest sIgA concentrations in the oldest elephant of their study (69 years of age). Another elephant in that study experienced a severe health event indicative of a systemic infection, and showed a four-fold increase in fecal IgA, suggesting it might be a useful health biomarker in elephants [24]. Previous studies in humans have shown decreases in sIgA associated with illnesses, such as upper respiratory tract infections [49] and malignant tumors [48]. Elevated sIgA concentrations also were found after administration of endotoxins to pigs [50] and dairy cows [51], indicating an immune response to pathogens. All elephants in the present study were checked by a veterinarian to ensure there were no underlying health conditions that could interfere with sIgA measures. Although saliva collection generally took less than 3 min, some elephants showed some agitation to the collector's hand swiping the inside the oral cavity. However, that was not reflected in significantly altered sIgA or cortisol concentrations.

Similar to previous studies, our results indicated a diurnal rhythm for salivary cortisol with peak concentrations in the morning (06:00 h) that gradually decreased throughout the day in a linear trend. Salivary cortisol was highest at 08:00 and lowest at 20:00 h in African elephants [31], which was similar to high values at 07:30 compared to 19:30 h for Asian elephants [30]. This is the same pattern observed for urine, where Brown et al. [29] reported a clear diurnal pattern of glucocorticoid excretion in Asian elephants, with the lowest concentrations observed just before midnight and peak concentrations occurring around 06:00–08:00 hours. However, in this study, clear patterns were not always observed during all collection periods. Five elephants exhibited more random cortisol fluctuations on one or more of the collection days, whereas only three showed the clear diurnal pattern on all three days. Various factors can disrupt normal patterns of cortisol, including stressful events during the day (social disputes, physical accidents, physical restraint) [31,52,53], age, sex, parturition, and environmental factors [52–55]. Casares et al. [31] revealed that the diurnal salivary cortisol pattern was disrupted by a fight between two zoo African elephants. The incident took place at 14:30 h and ended without human intervention, but the cortisol concentration of both individuals was increased two-fold at 16:00 h during the time it would normally have been declining. In our study, no obvious social disputes occurred between animals; however, the overall daily cortisol concentrations were highest on the first and lowest on the final day of collection, resulting in a statistically significant day effect. This suggests that the sample collection might have induced a mild stress response in some elephants, who then acclimated over time. Figure 2 shows that at 18:00 h, cortisol concentrations present the smallest variation. With low baseline concentrations and variations, sample collection is suggested during this time.

Some studies have reported significant correlations between sIgA and cortisol, including in humans [38] and dogs [56], while others found no such relationships. For example, Escribano et al. [22] revealed no significant correlation between the two parameters from pigs experiencing psychological stress in the form of isolation. Edwards et al. [24] also reported no correlation between salivary IgA and cortisol in Asian elephants from longitudinal samples over a six month period. In this study, no significant correlation between the two indicators was found, as evidenced by a linear downward trend in cortisol throughout the day, while sIgA tended to display a quartic pattern, with concentrations higher in the morning and evening.

5. Conclusions

For the first time, circadian effects on sIgA were evaluated in Asian elephants. This study revealed visible daily quartic trends of sIgA, providing basic knowledge of using sIgA as a biomarker for further studies. Results suggest that, just like for cortisol, time of day should be considered for saliva sample collection protocols for monitoring IgA. Moving forward, it will be important to understand differing response mechanisms when using IgA as a welfare indicator—chronic stressors may cause immune suppression and reductions in IgA, whereas acute illnesses could be associated with increases in IgA as part of an immune response to cope with underlying pathology. Thus, interpretation of IgA measures, just like GCs, may not always be straightforward. Both IgA and GCs have been shown to increase in response to acute stressors of a non-immune nature [13,57], and this certainly warrants further investigation before increased IgA concentrations can be considered a positive welfare indicator. As with other potential indicators of well-being, it is also important to understand normal physiological levels both within and between individuals, as well as in response to specific events. Biomarkers must be put into context, preferably by incorporating longitudinal measurements of multiple indicators, including IgA and GCs, to delineate concentrations indicative of an acute immune response or stressor, compared to those associated with longer-term positive or negative welfare states. The methodology described here provides a robust technique to investigate IgA in elephants, and these data provide a necessary baseline to interpret future data alongside other health and well-being measures, to determine whether incorporating IgA measurements will provide useful insight into elephant welfare.

Supplementary Materials: The following are available online at http://www.mdpi.com/2076-2615/10/1/157/s1, Figure S1: Individual IgA trends of all elephants, Figure S2: Individual cortisol trends of all elephants, Table S1: Raw data.

Author Contributions: Conceptualization, T.P., J.L.B., C.T., A.S.-F., and C.S.; Data curation, T.P., V.P.; Formal analysis, T.P., V.P., and C.S.; Funding acquisition, C.T. and A.S.F.; Investigation, T.P. and C.S.; Methodology, T.P., J.L.B., K.L.E., V.P., P.T., and C.S.; Project administration, T.P. and C.S.; Resources, C.T.; Software, V.P.; Supervision, J.L.B., A.S.-F., and C.S.; Validation, J.L.B., K.L.E., P.T. and C.S.; Visualization, T.P., J.L.B. and C.S.; Writing—original draft, T.P.; Writing—review and editing, T.P., J.L.B., C.T., A.S.-F., V.P. and C.S. All authors have read and agreed to the published version of the manuscript.

Funding: This research was funded by the Brian Nixon fund for the protection of elephants in Thailand and partially supported by Chiang Mai University.

Acknowledgments: The authors thank the elephant owners and mahouts for cooperating in this study and allowing us to work with the elephants. We are grateful to Jaruwan Khonmee, Pakkanut Bansiddhi, and Pallop Tankaew for providing laboratory assistance. A special thanks to Patiparn Toin, Wachiraporn Toonrongchang, Panida Muanghong, Sarisa Klinhom, Khajohnpat Boonprasert and Siripat Khammesi for assisting in sample collection throughout this study.

Conflicts of Interest: The authors declare no conflict of interest.

References

1. Bansiddhi, P.; Brown, J.L.; Thitaram, C.; Punyapornwithaya, V.; Somgird, C.; Edwards, K.L.; Nganvongpanit, K. Changing trends in elephant camp management in northern Thailand and implications for welfare. *PeerJ* **2018**, *6*, e5996. [CrossRef]
2. Möstl, E.; Palme, R. Hormones as indicators of stress. *Domest. Anim. Endocrinol.* **2002**, *23*, 67–74. [CrossRef]
3. Matteri, R.L.; Carroll, J.A.; Dyer, C.J. Neuroendocrine responses to stress. In *The biology of Animal Stress: Basic Principles and Implications for Animal Welfare*; Moberg, G.P., Mench, J.A., Eds.; CABI Publishing: Wallingford, UK, 2000; pp. 43–76.
4. Martin, P.A.; Crump, M.H. The adrenal gland. In *McDonald's Veterinary Endocrinology and Reproduction*; Iowa State Press: Wiley, IA, USA, 2003; pp. 165–200.
5. Menargues, A.; Urios, V.; Mauri, M. Welfare assessment of captive Asian elephants (*Elephas maximus*) and Indian rhinoceros (*Rhinoceros unicornis*) using salivary cortisol measurement. *Anim. Welf.* **2008**, *17*, 305–312.
6. Millspaugh, J.J.; Washburn, B.E. Use of fecal glucocorticoid metabolite measures in conservation biology research: Considerations for application and interpretation. *Gen. Comp. Endocrinol.* **2004**, *138*, 189–199. [CrossRef] [PubMed]
7. Sheriff, M.J.; Dantzer, B.; Delehanty, B.; Palme, R.; Boonstra, R. Measuring stress in wildlife: Techniques for quantifying glucocorticoids. *Oecologia* **2011**, *166*, 869–887. [CrossRef] [PubMed]
8. Bansiddhi, P.; Nganvongpanit, K.; Brown, J.L.; Punyapornwithaya, V.; Pongsopawijit, P.; Thitaram, C. Management factors affecting physical health and welfare of tourist camp elephants in Thailand. *PeerJ.* **2019**, *7*, e6756. [CrossRef] [PubMed]
9. Bansiddhi, P.; Brown, J.L.; Khonmee, J.; Norkaew, T.; Nganvongpanit, K.; Punyapornwithaya, V.; Angkawanish, T.; Somgird, C.; Thitaram, C. Management factors affecting adrenal glucocorticoid activity of tourist camp elephants in Thailand and implications for elephant welfare. *PLoS ONE* **2019**, *14*, e0221537. [CrossRef] [PubMed]
10. Whitham, J.C.; Wielebnowski, N. New directions for zoo animal welfare science. *Appl. Anim. Behav. Sci.* **2013**, *147*, 247–260. [CrossRef]
11. Yeates, J.W.; Main, D.C.J. Assessment of positive welfare: A review. *Vet. J.* **2008**, *175*, 293–300. [CrossRef]
12. Staley, M.; Conners, M.G.; Hall, K.; Miller, L.J. Linking stress and immunity: Immunoglobulin A as a non-invasive physiological biomarker in animal welfare studies. *Horm. Behav.* **2018**, *102*, 55–68. [CrossRef]
13. Tsujita, S.; Morimoto, K. Secretory IgA in saliva can be a useful stress marker. *Environ. Health Prev. Med.* **1999**, *4*, 1–8. [CrossRef] [PubMed]
14. Pressman, S.D.; Cohen, S. Does positive affect influence health? *Psychol. Bull.* **2005**, *131*, 925–971. [CrossRef] [PubMed]
15. Deinzer, R.; Kleineidam, C.; Stiller-Winkler, R.; Idel, H.; Bachg, D. Prolonged reduction of salivary immunoglobulin A (sIgA) after a major academic exam. *Int. J. Psychophysiol.* **2000**, *37*, 219–232. [CrossRef]
16. Takatsuji, K.; Sugimoto, Y.; Ishizaki, S.; Ozaki, Y.; Matsuyama, E.; Yamaguchi, Y. The effects of examination stress on salivary cortisol, immunoglobulin A, and chromogranin A in nursing students. *Biomed. Res.* **2008**, *29*, 221–224. [CrossRef]
17. Fan, Y.; Tang, Y.; Lu, Q.; Feng, S.; Yu, Q.; Sui, D.; Zhao, Q.; Ma, Y.; Li, S. Dynamic changes in salivary cortisol and secretory immunoglobulin A response to acute stress. *Stress Health* **2009**, *25*, 189–194. [CrossRef]
18. Kikkawa, A.; Uchida, Y.; Nakade, T.; Taguchi, K. Salivary secretory IgA concentrations in beagle dogs. *J. Vet. Med. Sci.* **2003**, *65*, 689–693. [CrossRef]
19. Svobodová, I.; Chaloupková, H.; Končel, R.; Bartoš, L.; Hradecká, L.; Jebavý, L. Cortisol and secretory immunoglobulin A response to stress in German shepherd dogs. *PLoS ONE* **2014**, *9*, e90820. [CrossRef]
20. Svobodová, I.; Vápeník, P.; Pinc, L.; Bartoš, L. Testing German shepherd puppies to assess their chances of certification. *Appl. Anim. Behav. Sci.* **2008**, *113*, 139–149. [CrossRef]
21. Muneta, Y.; Yoshikawa, T.; Minagawa, Y.; Shibahara, T.; Maeda, R.; Omata, Y. Salivary IgA as a useful non-invasive marker for restraint stress in pigs. *J. Vet. Med. Sci.* **2010**, *72*, 1295–1300. [CrossRef]
22. Escribano, D.; Gutiérrez, A.M.; Tecles, F.; Cerón, J.J. Changes in saliva biomarkers of stress and immunity in domestic pigs exposed to a psychosocial stressor. *Res. Vet. Sci.* **2015**, *102*, 38–44. [CrossRef]

23. Humphreys, A.F.; Tan, J.; Peng, R.; Benton, S.M.; Qin, X.; Worley, K.C.; Mikulski, R.L.; Chow, D.-C.; Palzkill, T.G.; Ling, P.D. Generation and characterization of antibodies against Asian elephant (*Elephas maximus*) IgG, IgM, and IgA. *PLoS ONE* **2015**, *10*, e0116318. [CrossRef] [PubMed]
24. Edwards, K.L.; Bansiddhi, P.; Paris, S.; Galloway, M.; Brown, J.L. The development of an immunoassay to measure immunoglobulin A in Asian elephant feces, saliva, urine and serum as a potential biomarker of well-being. *Conserv. Physiol.* **2019**, *7*. [CrossRef] [PubMed]
25. Hastings, M.H.; Maywood, E.S.; Brancaccio, M. Generation of circadian rhythms in the suprachiasmatic nucleus. *Nat. Rev. Neurosci.* **2018**, *19*, 453–469. [CrossRef] [PubMed]
26. Logan, R.W.; McClung, C.A. Rhythms of life: Circadian disruption and brain disorders across the lifespan. *Nat. Rev. Neurosci.* **2019**, *20*, 49–65. [CrossRef]
27. Shirakawa, T.; Mitome, M.; Oguchi, H. Circadian rhythms of S-IgA and cortisol in whole saliva—Compensatory mechanism of oral immune system for nocturnal fall of saliva. *Pediatric. Dent. J.* **2004**, *14*, 115–120. [CrossRef]
28. Heintz, M.R.; Santymire, R.M.; Parr, L.A.; Lonsdorf, E.V. Validation of a cortisol enzyme immunoassay and characterization of salivary cortisol circadian rhythm in chimpanzees (*Pan troglodytes*). *Am. J. Primatol.* **2011**, *73*, 903–908. [CrossRef]
29. Brown, J.L.; Kersey, D.C.; Freeman, E.W.; Wagener, T. Assessment of diurnal urinary cortisol excretion in Asian and African elephants using different endocrine methods. *Zoo Biol.* **2010**, *29*, 274–283. [CrossRef]
30. Menargues, A.; Urios, V.; Limiñana, R.; Mauri, M. Circadian rhythm of salivary cortisol in Asian elephants (*Elephas maximus*): A factor to consider during welfare assessment. *J. Appl. Anim. Welf. Sci.* **2012**, *15*, 383–390. [CrossRef]
31. Casares, M.; Silván, G.; Carbonell, M.D.; Gerique, C.; Martinez-Fernandez, L.; Cáceres, S.; Illera, J.C. Circadian rhythm of salivary cortisol secretion in female zoo-kept African elephants (*Loxodonta africana*). *Zoo Biol.* **2016**, *35*, 65–69. [CrossRef]
32. Hile, M.E.; Hintz, H.F.; Erb, H.N. Predicting Body Weight from Body Measurements in Asian Elephants (Elephas maximus). *J. Zoo Wildl. Med.* **1997**, *28*, 424–427.
33. Ng, V.; Koh, D.; Fu, Q.; Chia, S.-E. Effects of storage time on stability of salivary immunoglobulin A and lysozyme. *Clin. Chim. Acta* **2003**, *338*, 131–134. [CrossRef] [PubMed]
34. Team, R.C. *R: A Language and Environment for Statistical Computing*; R Foundation for Statistical Computing: Vienna, Austria, 2017.
35. Pinheiro, J.; Bates, D.; DebRoy, S.; Sarkar, D.; R. Core Team. *nlme: Linear and Nonlinear Mixed Effects Models*, R Package Version 3.1–143; 2019. Available online: https://cran.r-project.org/package=nlme (accessed on 12 December 2019).
36. Wickham, H. *ggplot2: Elegant Graphics for Data Analysis*; Springer: New York, NY, USA, 2016.
37. Bakdash, J.Z.; Marusich, L.R. Repeated Measures Correlation. *Front. Psychol.* **2017**, *8*, 456. [CrossRef] [PubMed]
38. Hucklebridge, F.; Clow, A.; Evans, P. The relationship between salivary secretory immunoglobulin A and cortisol: Neuroendocrine response to awakening and the diurnal cycle. *Int. J. Psychophysiol.* **1998**, *31*, 69–76. [CrossRef]
39. Dimitriou, L. Circadian effects on the acute responses of salivary cortisol and IgA in well trained swimmers. *Br. J. Sports Med.* **2002**, *36*, 260–264. [CrossRef]
40. Gleeson, M.; Cripps, A.W. Chapter 11—Ontogeny of mucosal immunity and aging. In *Mucosal Immunology*, 4th ed.; Mestecky, J., Strober, W., Russell, M.W., Kelsall, B.L., Cheroutre, H., Lambrecht, B.N., Eds.; Academic Press: Boston, MA, USA, 2015; pp. 161–185. ISBN 978-0-12-415847-4.
41. Park, S.-J.; Tokura, H. Bright light exposure during the daytime affects circadian rhythms of urinary melatonin and salivary immunoglobulin A. *Chronobiol. Int.* **1999**, *16*, 359–371. [CrossRef]
42. Mishra, K.P.; Yadav, A.P.; Ganju, L. Antarctic harsh environment as natural stress model: Impact on salivary immunoglobulins, transforming growth factor-β and cortisol level. *Indian J. Clin. Biochem.* **2012**, *27*, 357–362. [CrossRef]
43. Kugler, J.; Hess, M.; Haake, D. Secretion of salivary immunoglobulin a in relation to age, saliva flow, mood states, secretion of albumin, cortisol, and catecholamines in saliva. *J. Clin. Immunol.* **1992**, *12*, 45–49. [CrossRef]

44. Gleeson, M.; Cripps, A.W.; Clancy, R.L. Modifiers of the human mucosal immune system. *Immunol. Cell Biol.* **1995**, *73*, 397–404. [CrossRef]
45. Evans, P.; Der, G.; Ford, G.; Hucklebridge, F.; Hunt, K.; Lambert, S. Social class, sex, and age differences in mucosal immunity in a large community sample. *Brain Behav. Immun.* **2000**, *14*, 41–48. [CrossRef]
46. Corbett, L.; Muir, C.; Ludwa, I.A.; Yao, M.; Timmons, B.W.; Falk, B.; Klentrou, P. Correlates of mucosal immunity and upper respiratory tract infections in girls. *J. Pediatric. Endocrinol. Metab.* **2010**, *23*, 1–25. [CrossRef]
47. Jafarzadeh, A.; Sadeghi, M.; Karam, G.A.; Vazirinejad, R. Salivary IgA and IgE levels in healthy subjects: Relation to age and gender. *Braz. Oral Res.* **2010**, *24*, 21–27. [CrossRef] [PubMed]
48. Sun, H.; Chen, Y.; Zou, X.; Li, Q.; Li, H.; Shu, Y.; Li, X.; Li, W.; Han, L.; Ge, C. Salivary secretory immunoglobulin (SIgA) and lysozyme in malignant tumor patients. *BioMed Res. Int.* **2016**, *2016*. [CrossRef] [PubMed]
49. Orysiak, J.; Witek, K.; Zembron-Lacny, A.; Morawin, B.; Malczewska-Lenczowska, J.; Sitkowski, D. Mucosal immunity and upper respiratory tract infections during a 24-week competitive season in young ice hockey players. *J. Sports Sci.* **2017**, *35*, 1255–1263. [CrossRef] [PubMed]
50. Escribano, D.; Campos, P.H.R.F.; Gutiérrez, A.M.; Le Floc'h, N.; Cerón, J.J.; Merlot, E. Effect of repeated administration of lipopolysaccharide on inflammatory and stress markers in saliva of growing pigs. *Vet. J.* **2014**, *200*, 393–397. [CrossRef] [PubMed]
51. Iqbal, S.; Zebeli, Q.; Mansmann, D.A.; Dunn, S.M.; Ametaj, B.N. Oral administration of LPS and lipoteichoic acid prepartum modulated reactants of innate and humoral immunity in periparturient dairy cows. *Innate Immun.* **2014**, *20*, 390–400. [CrossRef] [PubMed]
52. Hellhammer, D.H.; Wüst, S.; Kudielka, B.M. Salivary cortisol as a biomarker in stress research. *Psychoneuroendocrinology* **2009**, *34*, 163–171. [CrossRef] [PubMed]
53. Gunnar, M.R.; Talge, N.M.; Herrera, A. Stressor paradigms in developmental studies: What does and does not work to produce mean increases in salivary cortisol. *Psychoneuroendocrinology* **2009**, *34*, 953–967. [CrossRef] [PubMed]
54. Behringer, V.; Clauß, W.; Hachenburger, K.; Kuchar, A.; Möstl, E.; Selzer, D. Effect of giving birth on the cortisol level in a bonobo groups' (*Pan paniscus*) saliva. *Primates* **2009**, *50*, 190–193. [CrossRef]
55. Palme, R. Non-invasive measurement of glucocorticoids: Advances and problems. *Physiol. Behav.* **2019**, *199*, 229–243. [CrossRef]
56. Skandakumar, S.; Stodulski, G.; Hau, J. Salivary IgA: A possible stress marker in dogs. *Anim. Welf.* **1995**, *4*, 339–350.
57. Jarillo-Luna, R.A.; Rivera-Aguilar, V.; Pacheco-Yépez, J.; Godínez-Victoria, M.; Oros-Pantoja, R.; Miliar-García, A.; Campos-Rodríguez, R. Nasal IgA secretion in a murine model of acute stress. The possible role of catecholamines. *J. Neuroimmunol.* **2015**, *278*, 223–231. [CrossRef] [PubMed]

Communication

Progesterone and Cortisol Levels in Blood and Hair of Wild Pregnant Red Deer (*Cervus Elaphus*) Hinds

Domenico Ventrella †, Alberto Elmi †, Martina Bertocchi *, Camilla Aniballi, Albamaria Parmeggiani, Nadia Govoni and Maria Laura Bacci

Department of Veterinary Medical Sciences, University of Bologna, 40064 Ozzano dell'Emilia (BO), Italy; domenico.ventrella2@unibo.it (D.V.); alberto.elmi2@unibo.it (A.E.); camilla.aniballi2@unibo.it (C.A.); albamari.parmeggiani@unibo.it (A.P.); nadia.govoni@unibo.it (N.G.); marialaura.bacci@unibo.it (M.L.B.)

* Correspondence: martina.bertocchi3@unibo.it

† Authors equally contributed.

Received: 26 November 2019; Accepted: 14 January 2020; Published: 16 January 2020

Simple Summary: The red deer, also known as the royal deer or European deer, is an artiodactyl mammal belonging to the Cervidae family, widely diffused in almost all of continental Europe. At the beginning of autumn in the Northern Hemisphere, the mating season begins. The males of red deer, called stags, are synchronized with the females, called hinds; indeed, at the beginning of the mating season, they show a marked increase in testosterone to match the hinds' estrus cycle. Gestation lasts about 230 days, so that calves are born in mid to late spring, the most favorable period for their survival. Scientific data on the reproduction physiology of this peculiar species in wild conditions are lacking, including hormonal variations during pregnancy. The present study describes mean levels of two critical hormones, cortisol and progesterone, in both blood and hair of wild pregnant red deer hinds. Correlation analysis confirmed how animals hunted in later phases of pregnancy have higher hair progesterone.

Abstract: The red deer (*Cervus elaphus* L., 1758) is one of the largest deer species in the world. Females are seasonal polyestrous, with negative photoperiod: the increase of the night peak of melatonin determines the secretion of GnRH and, therefore, LH and FSH. To date there is little information regarding the hormonal control during pregnancy for this species; this could be due to the difficulty of sampling wild subjects, while farmed animals' hormonal concentrations may not reflect the physiology of the animal in a natural state. In this study we evaluated the concentration of cortisol and progesterone, extracted from blood and hair, on 10 wild and pregnant red deer females. Belonging to the population of the Bolognese Apennines (Italy), the hinds were sampled in the January–March 2018 period, according to the regional selective hunting plan. Plasma progesterone (P4) ranged from a minimum of 1.9 to a maximum of 7.48 ng/mL; while hair P4 concentrations varied from 41.68 to 153.57 pg/mg. The plasma and hair cortisol ranges are respectively 0.4–2.97 ng/mL and 0.03–0.55 pg/mg; the only significant correlation was found between hair concentration of P4 and the date of death. The results of this preliminary study represent a small step towards a better knowledge of this species' physiology during pregnancy.

Keywords: red deer; hind; reproduction; progesterone; cortisol; hair

---

1. Introduction

The red deer (*Cervus elaphus* L., 1758) species, as the majority of the cervid taxa, shows a very deep relationship between its peculiar reproductive physiology and the their living environment [1]. It is widely diffused in almost all of continental Europe; in Italy it can be identified in a large Alpine area that extends from Cuneo to Udine, while in the Apennines the red deer occupies four distinct areas:

the first corresponds to most of the mountain territory of the provinces of Pistoia, Prato, Florence, and Bologna, the second to the Tuscan-Romagna Apennines, the third is represented by the Abruzzo National Park together with the neighboring territories and the fourth by the mountain massif of the Maiella. Lately, some specimens have also been found in southern regions, such as Sardinia [2]. The original habitat of the red deer is constituted by wooded areas with clearings or areas of little thick bush, generally in a flat environment or at low altitudes. However, over the years it has successfully adapted to live in different areas, from the heath to the forest of conifers, due to the increasing human competitive pressure [3].

In order to guarantee a good survival rate and growth of the offspring, hinds need to give birth at an appropriate time [4]. In the Mediterranean area, calves are usually born in spring and lactation goes on during summer [1], in order to exploit the most favorable time of the year, implying an extremely high synchronization between male and female reproductive activity during the rutting period. Red deer hinds are seasonally polyestrous, with a mean length of the estrous cycle of 18 ± 7 days and a gestation of approximately 231 days [5]. The reproductive cycle begins in October and if pregnancy does not occur, ovulation continues until March showing up to eight cycles [5]. The red deer, as well as fallow (*Dama dama*) and sika (*Cervus nippon*) deer, is a photoperiodic species: starting from the summer solstice, the decrease in daylight hours causes an increase in the duration of the nocturnal peak in melatonin secretion [6]. This event is responsible for the activation of the reproductive axis, leading to an increased secretion of LH releasing hormone, known as GnRH, and subsequent release of LH and FSH by the pituitary gland [7].

The afore-mentioned physiological events makes melatonin a major coordinator of the reproductive system and the annual rhythm [8]. The same endogenous circannual rhythm has been confirmed in red deer males, called stags, that show highly synchronized testicular cycles with a peak that overlaps the females' ovulation [5,9,10]. The almost complete suspension in spermatogenesis, during the non-rutting season (February–April), is further proof of such synchronization process [11]. Progesterone (P4) secreted by corpora lutea is necessary for the maintenance of pregnancy [12,13], while the contribution of placental progesterone is still to be assessed. P4 blood concentrations remain high (≅4 ng/mL) for the duration of gestation, and start decreasing one to two days before parturition in order to reach baseline levels after delivery [4,14].

In mammals, many other hormones undergo changes in blood concentration during the period of pregnancy, but how these relate to reproductive processes and their effects are poorly understood in wildlife. Cortisol (CORT) is another important hormone, generally used as stress marker also in wild animals [15], but the evaluation of its plasma concentrations is considered as poorly reliable since the hypothalamic–pituitary–adrenal axis is instantaneously activated upon stressful stimuli, such as restraint and blood sampling [16]. This is partially why, in the last years, alternative matrices for hormones' and other analytes' quantification have been proposed, including hair and feces, capable of providing different information regarding a longer timespan [17–19], also in red deer [20,21] and other frequently hunted ungulates [22]. For example, fecal CORT was proven to increase during late pregnancy with 10-year old females showing higher levels in comparison to five-year old hinds. It is interesting to note that fecal cortisol levels in wild animals increased during the calving period between May and June, potentially because of the stress related to delivery [23]. Despite the interest of the scientific community towards the physiology of reproduction of different species of cervids, there is currently only very few information regarding wild red deer hinds and pregnancy. Studies have been performed on farmed hinds [24], but such data could potentially differ from wildlife animals in light of the strong influences of the environment on such species. This lack of data can be partially imputed to the difficult nature of sampling. Since these animals can only be killed during certain periods of the year according to the local hunting regulation and usually undergo biometric examinations immediately after, collaborations with control centers for the collection of biometric records and hunters may help increasing the chances of sampling [19].

In light of the afore-mentioned reasons, the aim of this research was to analyze and describe hair and plasma levels of progesterone and cortisol from samples collected from pregnant wild red deer hinds killed during the hunting season in the south-western Bologna Apennines (Italy) area. Data were analyzed to highlight potential correlations with the age of the animals and the date of death. Hopefully, this will help gaining more knowledge regarding the reproduction of this wild species and its relationship with the environment.

2. Materials and Methods

2.1. Animals and Sampling

Ten (n = 10) pregnant red deer hinds were killed between 26 January and 1 March 2018 in the south-western Bologna Appennines (Italy) according to the regional hunting calendar (Resolution No. 473 of the Emilia Romagna Regional Executive, 10 April 2017). Upon death, hinds were immediately transferred to the pertinent control centers for the collection of biometric records where the samplings were performed. Ages were estimated from analyses of teeth eruption and wear patterns [25]. Pregnancy was confirmed by direct visualization of the uterus and identification of corpora lutea in the ovaries. Blood was collected from the jugular vein in sterile Lithium-heparin tubes, while hair was clipped from the dorsal-caudal region of the animals. All samples were refrigerated (5 ± 1 °C) and transferred in a cooler, within 24 h, to the physiology laboratories (ANFI-ASA) of the Department of Veterinary Medical Sciences of the University of Bologna (Ozzano dell'Emilia, Italy) as previously described [19].

2.2. Samples Preparation and Extraction

Upon arrival, blood samples (1 mL) were centrifuged (4000× *g*, 20 min) and stored at −20 °C until analysis for plasma hormone levels. Progesterone and cortisol were extracted by mixing 0.2 mL of plasma with 5 mL of petroleum ether or diethyl ether, respectively, for 30 min on a rotary mixer as described by Bono et al. [26]. The tubes were centrifuged at 2000× *g* for 15 min and the supernatants evaporated to dryness at 37 °C under an air-stream suction hood.

The hair samples were handled and analyzed as previously described [27]. Briefly, hair samples (250 mg) were washed with water and isopropanol in order to remove any organic residue from the surface. Once fully dried, samples were finely pulverized (180 mg) and incubated overnight with 6 mL of methanol for steroid extraction. After centrifugation, methanol was collected and evaporated to dryness under an air-stream suction hood.

2.3. Progesterone and Cortisol Radioimmunoassay

The dry extracts were stored at −20 °C until reconstitution in assay buffer for measurement of P4 (20 µL plasma equivalent, 6 mg hair equivalent) or cortisol (40 µL plasma equivalent, 60 mg hair equivalent) by radioimmunoassay; tritiated P4 (30 pg/0.1 mL; 96.6 Ci/mmol; PerkinElmer Inc., Boston, MA, USA) or tritiated cortisol (30 pg/0.1 mL; 94.6 Ci/mmol; PerkinElmer inc. Boston, MA, USA) were added, followed by rabbit anti-progesterone serum (0.1 mL, 1:10,000; antiserum produced in our laboratory) or rabbit anti-cortisol serum (0.1 mL, 1:20,000; produced in our laboratory), respectively. After incubation and separation of antibody-bound and -unbound steroid by charcoal-dextran solution (charcoal 0.25%, dextran 0.02% in phosphate buffer), tubes were centrifuged (15 min, 3000× *g*), the supernatant was decanted, and radioactivity immediately measured using a β-scintillation counter (Packard C1600, Perkin Elmer, USA).

The sensitivity of the P4 assay was 3.87 pg/tube, the intra- and inter-assay coefficients of variation were 5.7% and 9.4%, respectively. The sensitivity of the cortisol assay was 5.5 pg/tube, the intra- and inter-assay coefficients of variation were 4.9% and 8.7%, respectively.

Cross reactions of other steroids with antiserum raised against P4 were progesterone (100%), 11α-hydroxyprogesterone (9.7%), 5α-pregnan-3-20-dione (4.4%), 17α-hydroxyprogesterone (1.5%), 20α-hydroxyprogesterone (0.3%), cortisol (0.05%), testosterone, and 17β-estradiol (<0.001%).

Cross reactions of various steroids with antiserum raised against cortisol were cortisol (100%), cortisone (5.3%), 11α-deoxycortisol (5.0%), corticosterone (9.5%), 20α-dihydrocortisone (0.4%), prednisolone (4.60%), progesterone, and testosterone (<0.001%).

In order to determine the parallelism between hormone standards and endogenous hormones in red deer hair, a pool sample containing high concentrations of cortisol and progesterone was serially diluted (1:1–1:8) with assay buffer. A regression analysis was used to determine parallelism between the two hormone levels in the same assay. A high degree of parallelism was confirmed by regression test ($R^2 = 0.98$, $p < 0.01$).

The assay results for both hormones were expressed as ng/mL for plasma and as pg/mg for hair.

2.4. Data Analysis

Descriptive statistics were calculated and expressed as means, standard deviations, and min/max values using the MedCalc Statistical Software version 18.11.3 (MedCalc Software bvba, Ostend, Belgium; https://www.medcalc.org; 2019). In order to evaluate any potential relationships between the analyzed parameters, a non-parametric Spearman rank correlation test was performed. The statistical significance was set at $p < 0.05$ (95% C.I.).

3. Results

The results of the descriptive statistics are summarized in Table 1. All the animals were killed between the end of January and the first of March, therefore, according to the reproductive physiology of the species, hinds were approximately at the third to fifth month of gestation. Age ranged from approximately two up to five years. Plasma levels of progesterone ranged from 1.9 to 7.48 ng/mL with a mean of 3.86 ng/mL (SD = 2.14), while plasma cortisol ranged between 0.4 ng/mL and 2.97 ng/mL (mean = 1.44 ng/mL SD = 0.96). On the other hand, the ranges of hair cortisol and P4 were respectively 0.03–0.55 pg/mg (mean = 0.29 pg/mg) and 41.68–153.57 pg/mg (mean = 73.05 pg/mg).

Table 1. Descriptive data of the red deer hinds and quantified hormones.

Animal	Date of Death	Age (Months)	Plasma (ng/mL)		Hair (pg/mg)	
			P4	CORT	P4	CORT
1	26 January	36	3.23	0.77	19.47	0.50
2	9 February	22	2.20	0.38	41.68	0.05
3	12 February	45	3.15	1.87	50.76	0.23
4	19 February	30	3.17	2.97	96.91	0.16
5	19 February	32	7.44	1.79	153.57	0.36
6	19 February	46	7.48	1.91	31.40	0.21
7	19 February	36	5.44	2.73	69.52	0.55
8	1 March	31	2.51	0.50	104.75	0.51
9	1 March	43	2.09	0.40	80.13	0.03
10	1 March	58	1.90	1.05	82.3	0.28
mean (SD)		37.9 (10.3)	3.86 (2.14)	1.44 (0.96)	73.05 (39.85)	0.29 (0.19)
min–max		22–58	1.90–7.48	0.38–2.97	19.47–153.57	0.03–0.55

P4: progesterone; CORT: cortisol.

The Spearman rank correlation analysis, reported in Table 2, did not show any significant result aside from a significant correlation ($p = 0.042$) between hair concentration of P4 and the date of death, and a weak correlation, not statistically significant, between plasma concentrations of P4 and CORT.

Table 2. Spearman rank correlation coefficients (ρ) table with *p* values (95% C.I.).

	Plasma CORT	Hair CORT	Plasma P4	Hair P4	Age	Date of Death
Plasma CORT		0.236 $p = 0.511$	0.612 $p = 0.060$	0.091 $p = 0.809$	0.182 $p = 0.614$	−0.051 $p = 0.897$
Hair CORT			0.358 $p = 0.310$	0.152 $p = 0.676$	0.030 $p = 0.934$	0.006 $p = 0.996$
Plasma P4				−0.164 $p = 0.652$	−0.036 $p = 0.920$	−0.380 $p = 0.277$
Hair P4					−0.261 $p = 0.466$	0.659 $p = 0.042$
Age						0.276 $p = 0.436$
Date of Death						

Date of Death was reported according to the Julian calendar.

4. Discussion

As stated in the introduction section, one of the main limitations in studying wildlife reproductive physiology lies within the nature of the sampling possibilities. It is indeed almost impossible to collect repeated samples on the same animal in order to obtain a dynamic profile of specific analytes throughout the different phases of gestation. This is why collecting samples upon killing during the hunting season may represent a good, if not the only, way to perform this kind of research [19]. This study confirms the chance to collaborate with hunters and local biometric centers.

It is widely recognized that analyzing hormones in hair provides information regarding a longer timespan when compared to blood analyses [19,20], but being able to specifically identify such timespan can be challenging. Indeed, due to the nature of the sampling, it is not always possible to completely trim the chosen area to calculate the growth rate of the hair in wildlife animals prior to a subsequent hair sampling. The red deer undergoes two pelages, also known as moultings, one in spring and one in autumn, approximately in May and October, respectively [28,29]. Therefore, the hair collected in the present study belongs to the winter coat and should provide info regarding few months before trimming.

The hormones analyzed in the present study, in both matrices, showed wide ranges suggesting strong individual variations. The differences in progesterone levels may be related to the fact that the sampling took place throughout more than one month and that hinds may be in different phases of pregnancy (approximately from the third up to the fifth month of gestation). Moreover, it was already proved that, in this species, the levels of plasma progesterone fluctuate and correlate with the number of corpora lutea present [14]. On the other hand, cortisol variations, already described in literature for the red deer [20], are related to a multitude of environmental, seasonal, and physiological factors.

The mean level of blood progesterone recorded in the present study (3.86 ng/mL) is in accordance with previous studies performed in pregnant red deer hinds farmed in paddocks [14,30]. Such agreement of data seems to support the possibility to use data collected from farmed animals also for wildlife. Such statement is extremely preliminary and particular attention will have to be paid if the analytes were to concern stress patterns, which will necessarily be influenced by husbandry conditions and environmental factors. Overall, relatively higher levels of plasma progesterone have to be expected in pregnant hinds when compared to non-pregnant animals, as already reported [31].

The mean hair progesterone level was 73.05 pg/mg with high individual variations most likely explained by the variations detected in plasma levels. As of today, to the best of the authors' knowledge, no studies were published assessing hair levels of this hormone in red deer and cervids in general. It is therefore quite challenging to discuss such findings. What can be said, on the basis of the present study, is that the analytical approach seems to be reliable and relatively easy to perform. Hair progesterone was assessed in cows in a recent study were no differences between pregnant and non-pregnant animals were found [32]. The authors hypothesized that the cause behind the lack of differences may be the short period between calving and the next successful insemination [32]. Again, it is impossible to compare the results with the ones obtained in the present study in light of the physiological differences between the species and the zootechnical pressure dictated by industry demands.

When compared to progesterone, cortisol is overall more influenced by the conditions in which the animals live [15]. Blood cortisol levels change quickly throughout the day and are therefore relatively unreliable, but quantification in matrices such as feces and hair seems to be a good indicator of chronic stress, potentially also related to the reproductive status of the animal [16,23]. Moreover, cortisol, in all matrices, shows high seasonality variations in red deer [33] as in other wild ungulates [34].

Our results show, as expected, highly variable plasma CORT levels ranging from 0.38 to 2.97 ng/mL. This is likely to be imputed to several "acute" factors and the variability of this hormone pattern itself [22,34]. On the other hand, hair CORT levels are the reflection of the continuous incorporation of the hormone into the hair shaft [15], and thus less influenced by acute stressors. This is why this matrix has been addressed in the last years as highly efficient in wildlife [19]. A study performed on Italian red deer reported mean values of hair CORT of 5.75 pg/mg in females [20], almost 20 times higher than the mean value obtained in the present study (0.29 pg/mg). Factors influencing cortisol production during pregnancy are extremely variable and partially unknown [35]. What is known is that cortisol increases in the late phase of a pregnancy [23], and that, in cows, parturition in the month preceding sampling increases hair cortisol levels [36]. Animals sampled in the present study were approximately at the third month of gestation, and therefore in the first phase of pregnancy. It is possible to suppose that hair levels of CORT would have been higher in case of later phases of gestation. The correlation between hair CORT and chronic rather than acute stress was further confirmed by a study carried out on a red deer population from the central Italian Alps, Sondrio Province [20].

Our correlation analysis did not highlight any relationship between the analyzed parameters aside from a significant correlation ($\rho = 0.659$; $p = 0.042$) between hair P4 and the date of death. This correlation may be explained by the ongoing pregnancy, characterized by high levels of P4. Therefore, it is likely that hinds killed later on have had higher accumulation of this hormone in the hair. A weak correlation, not statistically significant, was also found between the plasma levels of the two hormones, P4 and CORT ($\rho = 0.612$; $p = 0.060$). Nonetheless, it is important to acknowledge that the low sample size may be responsible for a weak power of the statistical analysis.

The present study strengthens the use of hair for endocrinological evaluations in wild animals, in light of the easy and non-invasive sampling procedure and the representation of longer time periods in one single sample [15]. Our purpose was to describe the concentration of progesterone and cortisol, from hair and blood, of the wild population of red deer females (*Cervus elaphus* L., 1758) living in the Apennine area, during a particular physiological situation such as gestation. In conclusion, performing such studies on hunted animals seems to be useful in better understanding the physiology of wildlife animals and may provide new useful data that can benefit the species and the scientific community.

Author Contributions: A.E. and M.L.B. conceptualized the study; A.E. coordinated the sampling procedures; M.B. and N.G. performed the analyses; A.E. and D.V. analyzed the data; D.V. and C.A. drafted the manuscript; N.G., M.L.B., and A.P. revised and approved the final draft. All authors have read and agreed to the published version of the manuscript.

Funding: The research was funded by RFO-UNIBO.

Acknowledgments: Authors would like to thank the volunteers of "URCA Bologna" Association and the "ATC BO 3" for their help in the sampling process and Stefano Mattioli for the age estimations.

Conflicts of Interest: The authors declare no conflict of interest.

References

1. Peláez, M.; San Miguel, A.; Rodríguez-Vigal, C.; Perea, R. Climate, female traits and population features as drivers of breeding timing in Mediterranean red deer populations. *Integr. Zool.* **2017**, *12*, 396–408. [CrossRef]

2. Zachos, F.E.; Frantz, A.C.; Kuehn, R.; Bertouille, S.; Colyn, M.; Niedziałkowska, M.; Pérez-González, J.; Skog, A.; Sprěm, N.; Flamand, M.-C. Genetic Structure and Effective Population Sizes in European Red Deer (*Cervus elaphus*) at a Continental Scale: Insights from Microsatellite DNA. *J. Hered.* **2016**, *107*, 318–326. [CrossRef]
3. Minelli, A. *La Fauna in Italia*; Touring Editore: Milan, Italy, 2002; ISBN 978-88-365-2621-5.
4. Asher, G.W. Reproductive cycles of deer. *Anim. Reprod. Sci.* **2011**, *124*, 170–175. [CrossRef]
5. Guinness, F.; Lincoln, G.A.; Short, R.V. The reproductive cycle of the female red deer, *Cervus elaphus* L. *J. Reprod. Fertil.* **1971**, *27*, 427–438. [CrossRef] [PubMed]
6. Lincoln, G.A. Biology of seasonal breeding in deer. In *The Biology of Deer*; Springer Science & Business Media Press: Berlin, Germany, 1992; pp. 565–574.
7. Lincoln, D.W.; Fraser, H.M.; Lincoln, G.A.; Martin, G.B.; Mcneilly, A.S. Hypothalamic pulse generators. In *Proceedings of the Proceedings of the 1984 Laurentian Hormone Conference*; Greep, R.O., Ed.; Academic Press: Cambridge, MA, USA, 1985; pp. 369–419.
8. Tamarkin, L.; Baird, C.J.; Almeida, O.F. Melatonin: A coordinating signal for mammalian reproduction? *Science* **1985**, *227*, 714–720. [CrossRef] [PubMed]
9. Mitchell, B.; Lincoln, G.A. Conception dates in relation to age and condition in two populations of red deer in Scotland. *J. Zool.* **1973**, *171*, 141–152. [CrossRef]
10. Leader-Williams, N. *Reindeer on South Georgia: the Ecology of an Introduced Population*; Cambridge University Press: Cambridge, UK, 1988.
11. Hochereau-de Reviers, M.T.; Lincoln, G.A. Seasonal variation in the histology of the testis of the red deer, *Cervus elaphus*. *J. Reprod. Fertil.* **1978**, *54*, 209–213. [CrossRef] [PubMed]
12. Asher, G.W.; Fisher, M.W.; Berg, D.K.; Waldrup, K.A.; Pearse, A.J. Luteal support of pregnancy in red deer (*Cervus elaphus*): Effect of cloprostenol, ovariectomy and lutectomy on the viability of the post-implantation embryo. *Anim. Reprod. Sci.* **1996**, *41*, 141–151. [CrossRef]
13. Plotka, E.D.; Seal, U.S.; Verme, L.J.; Ozoga, J.J. Reproductive steroids in white-tailed deer. IV. Origin of progesterone during pregnancy. *Biol. Reprod.* **1982**, *26*, 258–262. [CrossRef]
14. Kelly, R.W.; McNatty, K.P.; Moore, G.H.; Ross, D.; Gibb, M. Plasma concentrations of LH, prolactin, oestradiol and progesterone in female red deer (*Cervus elaphus*) during pregnancy. *Reproduction* **1982**, *64*, 475–483. [CrossRef]
15. Heimbürge, S.; Kanitz, E.; Otten, W. The use of hair cortisol for the assessment of stress in animals. *Gen. Comp. Endocrinol.* **2019**, *270*, 10–17. [CrossRef] [PubMed]
16. Davenport, M.D.; Tiefenbacher, S.; Lutz, C.K.; Novak, M.A.; Meyer, J.S. Analysis of endogenous cortisol concentrations in the hair of rhesus macaques. *Gen. Comp. Endocrinol.* **2006**, *147*, 255–261. [CrossRef] [PubMed]
17. Sheriff, M.J.; Dantzer, B.; Delehanty, B.; Palme, R.; Boonstra, R. Measuring stress in wildlife: Techniques for quantifying glucocorticoids. *Oecologia* **2011**, *166*, 869–887. [CrossRef] [PubMed]
18. Narayan, E.J. Evaluation of physiological stress in Australian wildlife: embracing pioneering and current knowledge as a guide to future research directions. *Gen. Comp. Endocrinol.* **2017**, *244*, 30–39. [CrossRef] [PubMed]
19. Ventrella, D.; Elmi, A.; Barone, F.; Carnevali, G.; Govoni, N.; Bacci, M.L. Hair Testosterone and Cortisol Concentrations in Pre- and Post-Rut Roe Deer Bucks: Correlations with Blood Levels and Testicular Morphometric Parameters. *Animals* **2018**, *8*, 113. [CrossRef]
20. Caslini, C.; Comin, A.; Peric, T.; Prandi, A.; Pedrotti, L.; Mattiello, S. Use of hair cortisol analysis for comparing population status in wild red deer (*Cervus elaphus*) living in areas with different characteristics. *Eur. J. Wildl. Res.* **2016**, *62*, 713–723. [CrossRef]
21. Montillo, M.; Caslini, C.; Peric, T.; Prandi, A.; Netto, P.; Tubaro, F.; Pedrotti, L.; Bianchi, A.; Mattiello, S. Analysis of 19 Minerals and Cortisol in Red Deer Hair in Two Different Areas of the Stelvio National Park: A Preliminary Study. *Animals* **2019**, *9*, 492. [CrossRef]
22. Pecorella, I.; Ferretti, F.; Sforzi, A.; Macchi, E. Effects of culling on vigilance behaviour and endogenous stress response of female fallow deer. *Wildl. Res.* **2016**, *43*, 189–196. [CrossRef]
23. Pavitt, A.T.; Pemberton, J.M.; Kruuk, L.E.B.; Walling, C.A. Testosterone and cortisol concentrations vary with reproductive status in wild female red deer. *Ecol. Evol.* **2016**, *6*, 1163–1172. [CrossRef]

24. Patel, K.K.; Howe, L.; Heuer, C.; Asher, G.W.; Wilson, P.R. Pregnancy and mid-term abortion rates in farmed red deer in New Zealand. *Anim. Reprod. Sci.* **2018**, *193*, 140–152. [CrossRef]
25. Mattioli, S.; De Marinis, A.M. 2009–Guida al rilevamento biometrico degli Ungulati. Istituto Su-periore per la Protezione e la Ricerca Ambientale, Documenti Tecnici. Volume 28, pp. 1–216. Available online: http://www.isprambiente.gov.it/files/pubblicazioni/guida-rilevamento-ungulati.pdf (accessed on 15 October 2019).
26. Bono, G.; Cairoli, F.; Tamanini, C.; Abrate, L. Progesterone, estrogen, LH, FSH and PRL concentrations in plasma during the estrous cycle in goat. *Reprod. Nutr. Dev.* **1983**, *23*, 217–222. [CrossRef] [PubMed]
27. Bacci, M.L.; Nannoni, E.; Govoni, N.; Zannoni, A.; Forni, M.; Martelli, G.; Sardi, L. Hair cortisol level determination in sows during successive reproductive cycles in different seasons. *Reprod. Biol.* **2014**, *14*, 218–223. [CrossRef] [PubMed]
28. Ryder, M.L.; Kay, R.N.B. Structure of and seasonal change in the coat of Red deer (*Cervus elaphus*). *J. Zool.* **1973**, *170*, 69–77. [CrossRef]
29. Loudon, A.S.; Milne, J.A.; Curlewis, J.D.; McNeilly, A.S. A comparison of the seasonal hormone changes and patterns of growth, voluntary food intake and reproduction in juvenile and adult red deer (*Cervus elaphus*) and Père David's deer (*Elaphurus davidianus*) hinds. *J. Endocrinol.* **1989**, *122*, 733–745. [CrossRef] [PubMed]
30. Adam, C.L.; Moir, C.E.; Atkinson, T. Plasma concentrations of progesterone in female red deer (*Cervus elaphus*) during the breeding season, pregnancy and anoestrus. *Reproduction* **1985**, *74*, 631–636. [CrossRef]
31. Korzekwa, A.J.; Szczepańska, A.; Bogdaszewski, M.; Nadolski, P.; Malż, P.; Giżejewski, Z. Production of prostaglandins in placentae and corpus luteum in pregnant hinds of red deer (*Cervus elaphus*). *Theriogenology* **2016**, *85*, 762–768. [CrossRef] [PubMed]
32. Tallo-Parra, O.; Carbajal, A.; Monclús, L.; Manteca, X.; Lopez-Bejar, M. Hair cortisol and progesterone detection in dairy cattle: interrelation with physiological status and milk production. *Domest. Anim. Endocrinol.* **2018**, *64*, 1–8. [CrossRef] [PubMed]
33. Ingram, J.R.; Crockford, J.N.; Matthews, L.R. Ultradian, circadian and seasonal rhythms in cortisol secretion and adrenal responsiveness to ACTH and yarding in unrestrained red deer (*Cervus elaphus*) stags. *J. Endocrinol.* **1999**, *162*, 289–300. [CrossRef]
34. Fattorini, N.; Lovari, S.; Brunetti, C.; Baruzzi, C.; Cotza, A.; Macchi, E.; Pagliarella, M.C.; Ferretti, F. Age, seasonality, and correlates of aggression in female Apennine chamois. *Behav. Ecol. Sociobiol.* **2018**, *72*, 171. [CrossRef]
35. Bleker, L.S.; Roseboom, T.J.; Vrijkotte, T.G.; Reynolds, R.M.; de Rooij, S.R. Determinants of cortisol during pregnancy–The ABCD cohort. *Psychoneuroendocrinology* **2017**, *83*, 172–181. [CrossRef]
36. Braun, U.; Michel, N.; Baumgartner, M.R.; Hässig, M.; Binz, T.M. Cortisol concentration of regrown hair and hair from a previously unshorn area in dairy cows. *Res. Vet. Sci.* **2017**, *114*, 412–415. [CrossRef] [PubMed]

Article

Development and Implementation of Baseline Welfare Assessment Protocol for Captive Breeding of Wild Ungulate—Punjab Urial (*Ovis vignei punjabiensis*, Lydekker 1913)

Romaan Hayat Khattak, Zhensheng Liu * and Liwei Teng

College of Wildlife and Protected Area, Northeast Forestry University, Harbin 150040, China; romaanktk@gmail.com (R.H.K.); tenglw1975@163.com (L.T.)

* Correspondence: zhenshengliu@163.com

Received: 25 October 2019; Accepted: 5 December 2019; Published: 9 December 2019

Simple Summary: Current issues concerning animal welfare are receiving remarkable consideration among scientific communities and wildlife conservation organizations. In addition to the ethical and legal grounds of animal welfare, it is important to produce healthy and viable populations in captivity by ensuring optimum welfare. Wild species are reared in captivity but, unlike domestic animals, there is a lack of welfare assessment protocols for these wild species. In the current study, we developed and applied the first baseline welfare assessment protocol for Punjab urial (*Ovis vignei punjabiensis*). While developing this protocol, we gathered all possible and existing information about the biology and ecology of this species in its natural habitat. Since no welfare assessment protocol and husbandry guidelines exist for this species, we used a welfare assessment protocol for domestic sheep as the base and reference in developing the current protocol for captive Punjab urial. Later, we applied this protocol to three different herds of Punjab urial at two different facilities. Based on the initial results obtained, some areas were found to have shortcomings and recommended for quick corrective measures and improvements.

Abstract: To ensure that captive breeding and other associated programs are more robust and sustainable, it is of utmost importance to ensure optimum welfare. Although it is well known that standard welfare is crucial for successful captive breeding, there is still a lack of welfare assessment protocols for wild species. The current study aimed to develop a leading baseline welfare assessment protocol for assessing welfare in captive Punjab urial. This protocol is based on the welfare protocol for domestic sheep from the Welfare Quality® project, coupled with all the information obtained from the published literature on the species' biology and ecology. This protocol consists of 4 principles, 12 criteria, and 31 animal- and resource-based indicators. The protocol was tested and applied to three different herds of Punjab urial at two different facilities. Initial results showed that some areas need to be improved for better captive breeding and management.

Keywords: welfare; protocol; veterinary assessment; Punjab urial; body condition; behavior; Pakistan

---

1. Introduction

To ensure that captive breeding programs are more robust and sustainable, it is important to ethically and legally ensure optimum animal welfare [1]. The rapid developmental processes of humans are affecting the natural habitats of wildlife. Thus, wildlife reservoirs, zoos, and enclosures must be adapted to minimize the effect of these changes [2]. Today, captive breeding is one of the most important conservation tools [3], providing an opportunity to the rare endangered species to produce stable populations for eventual release into the wild [4].

In the meantime, enclosures and other facilities for wild animals are under severe pressure to limit animals to small areas [5]. Thus, animal welfare subjects, particularly those related to the captive wild species, are rapidly recognized [6,7]. It is important to ensure the optimum levels of animal welfare of captive animals for the production and maintenance of healthy, viable populations [8]. To make captive breeding more robust, it is important to determine the main factors of species welfare and, most importantly, the welfare of every individual of a particular group [9].

Animal welfare assessment protocols can be designed by gathering information through simple inspections, animal observations, and visiting animal facilities and enclosures [10]. Animal welfare assessment protocols for livestock (poultry, cattle, and pigs) have already been developed under the auspices of the Welfare Quality® project. These protocols are mainly based on animal-based indicators, in addition to also having resource or environmental measures [11]. Animal-based measures can be directly recorded by observing animals, including their physical appearance, health, and behavior. Unlike animal-based measures, environmental measures assess the available resources for these animals in captivity, and the animal itself is not taken into account.

Ovis orientalis (urial) is a wild sheep that closely resembles Marco Polo sheep in general body appearance [12]. In Pakistan, the urial is represented by three subspecies: *Ovis vignei vigeni* (Ladakh urial), which is restricted to northern areas (Gilgit-Baltistan) of Pakistan; *Ovis vignei punjabiensis* (Punjab urial) found in the Salt Range (Punjab) and the northwestern province of Khyber Pakhtunkhwa (formerly known as the North-West Frontier Province); and *Ovis vignei blanfordi* (Baluchi urial), which is found in the southwestern province of Balochistan [13]. The Punjab urial is a gregarious ungulate, and most big herds include females, lambs, and immature males. It has been observed that the urial generally prefers grasses, but can also be found foraging on shrubs [14].

The species has been declared as vulnerable globally, with a declining population trend, according to the International Union for Conservation (IUCN) list of threatened species [15], and is endangered in Pakistan [16–18]. In Pakistan, wild ungulates are reared in captivity, but there are no standardized methods and protocols to measure the welfare of these captive ungulates. Therefore, this current study aimed to (i) design and develop a baseline welfare assessment protocol for captive Punjab urial based on the livestock welfare assessment protocol from the Welfare Quality® project, (ii) implement this welfare assessment protocol in facilities hosting Punjab urial, and (iii) suggest recommendations for better captive breeding and management.

2. Materials and Methods

For this study, we selected the subspecies *Ovis orientalis punjabiensis* because of its availability in captivity. The current study was conducted in two steps. In the first step, the welfare assessment protocol was developed, and, in the second step, the newly established protocol was implemented at captive facilities housing Punjab urial. We developed the welfare assessment protocol by combining results from other studies on the biology and behavior of the species and the sheep welfare assessment protocol [19], as both domestic sheep and Punjab urial belong to the family Bovidae.

To obtain information on the general biology and behavior of the species, we used Google Scholar and Web of Science search engines using "Ovis vignei" as keywords. Limited scientific published information is available regarding the biology of this species in natural habitats. Moreover, no work has been conducted to investigate the problems faced by this species in captivity. More than 31 scientific published papers were reviewed. Most of these published work focused on population status, population dynamics, diet ecology, and habitat, of which two papers [12,20] provided useful detailed information on the behavior and general biology of the species that was utilized for the welfare assessment protocol.

For developing the Punjab urial welfare assessment protocol, four basic principles were taken into account: good feeding, good housing, good health, and suitable behavior [10,11]. These principles led to twelve criteria (see Section 3), which in turn allowed the development of welfare assessment indicators [21]. After combining information from [19,22], we developed an extensive set of 31 welfare indicators for Punjab urial (see Section 3).

The final version of the welfare assessment protocol was then applied to three different groups of captive Punjab urial at two different facilities—Cherat Wildlife Park (CWP) in Nowshera and Manglot Wildlife Park (MWP) in Nizampur—in the month of August 2019. Both of these parks are located in the Nowshera District of Khyber Pakhtunkhwa Province, Pakistan. The captive breeding program was launched in 2008 at CWP with a single pair of founder animals, while in 2012 it was initiated at MWP, also with a single pair. The groups of both programs were mixed herds, including adult males, sub adult males, adult females, sub adult females, and lambs. CWP had one group consisting of 23 individuals ($n = 23$) with a mean age of 3.21 ± 2.21 years. At MWP, two groups were present ($n = 6$ and $n = 8$), with mean ages of 3.16 ± 1.57 and 3.33 ± 1.69 years, respectively. Groups were represented by coding their facilities (centers) as captive Punjab urial-1 (CU1) at CWP, and captive Punjab urial-2 (CU2) and captive Punjab urial-3 (CU3) at MWP. The protocol was applied by the same person.

Statistical Analysis

In the current protocol, the observation time for social behavior was 120 min per herd in six 20-min sessions. Continuous focal sampling was used because these behaviors may be of short duration and elusive nature [10]. We performed the Shapiro–Wilk test with our datasets to check for normality. Later, data recorded for social behavior (n = number of occurrences of an event) for all three groups in the current protocol were analyzed using the nonparametric Kruskal–Wallis test to determine the significance of observed variables. All the statistical analyses were performed using SPSS version 23.0 (IBM Corp., Armonk, NY, USA).

3. Results

3.1. Development of Welfare Assessment Protocol

The protocols developed in the current study for welfare assessment of Punjab urial consisted of 4 basic principles, 12 criteria, and 31 indicators (Table 1). Of the 12 criteria of the Welfare Quality® protocol for domestic sheep, some were replaced keeping in view the biology of the species. The species under study is a wild ungulate, thus "lack of minerals" was added to the list of criteria. The current study is a leading step in captive Punjab urial welfare assessment, and will provide a base to investigate more complex and less frequently studied aspects in captive animals.

Table 1. Punjab urial welfare quality assessment protocol principles, criteria, and indicators. Animal-based indicators are represented by *.

Welfare Principles	Criteria	Indicators
Good feeding	1. Lack of prolonged appetite	1.1 Body conditions *
	2. Lack of prolonged thirst	2.1 Number of water sources 2.2 Water availability 2.3 Cleanliness of water sources
	3. Lack of minerals	3.1 Availability of salt licks 3.2 Licking objects
Good housing	4. Thermal ease	4.1 Shelter availability 4.2 Shade availability
	5. Easiness in movement	5.1 Total area of enclosure 5.2 Space (m^2) offered per animal
	6. Standard enclosures	6.1 Fence conditions 6.2 Fence substratum 6.3 Availability of quarantine 6.4 Number of quarantines
Good health	7. Lack of injuries	7.1 Integument deformities * 7.2 Lameness *
	8. Lack of disease	8.1 Ophthalmic discharge * 8.2 Nasal discharge * 8.3 Labored breathing * 8.4 Diarrhea * 8.5 Availability of veterinarian 8.6 Availability of veterinary facility
Appropriate behavior	9. Displaying social behavior	9.1 Affinitive interactions * 9.2 Agonistic interactions *
	10. Group dynamics	10.1 Herd size 10.2 Herd composition 10.3 Number of animals (other species)
	11. Display of other behavior	11.1 Stereotypic behavior * 11.2 Environmental enrichment programs
	12.Good human–animal affiliations	12.1 Medical training program 12.2 Capturing, handling, immobilization, and translocation

3.1.1. Lack of Prolonged Appetite

Since there are no husbandry guidelines for this species, there is no parameter that can be used for the description of this criterion. In many animal welfare assessment protocols, however, body condition is included as an animal-based indicator to measure welfare. Both poor and excessive body conditions can lead to extreme and unexpected results. Poor body conditions can show signs of malnutrition, diseases, or extreme hunger. Poor and weak body conditions can have severe negative effects on the reproduction, behavior, and health of the animals [23].

To indicate the welfare problems, both poor and excessive body conditions can be used (indicator 1.1) [24]. To date, there is no scoring scale for the body condition of Punjab urial; hence, the body condition scoring guidelines for two other wild bovids proposed in [10,24] were followed. Animals were observed from a close distance from behind and the side. Punjab urial was scored as "poor body conditions" if ribs, spine, and pelvis were sharp and prominent, and projected with a concave rump. The animal was scored as "normal body condition" if ribs, pelvis, and spine were difficult to be distinguished from one another, with a flat rump area. The animal was scored as "excessive body conditions" if ribs, spine, and pelvis were not visible because of a thick layer of fat, with a protruding rump area.

3.1.2. Lack of Prolonged Thirst

One of the most important welfare requirements is ad libitum, access to water. Water provision in animal welfare protocol is a resource-based indicator. There should be easy access to water sources for the animals, containing clean and fresh water supplied on a daily basis.

According to the current study, water availability (indicator 2.1), in conjunction with the number of water sources (2.2) and cleanliness of water sources (2.3), should be properly checked. All these indicators were validated in [10,25] for assessing the welfare of captive *Dorcas gazelles* and cows. Water was scored as "dirty" if it was polluted with food wastes or plastics, or was dark green in color with a foul smell. On the contrary, water was scored as "clean" if it was without any contamination or foul smell. Remains of fresh feed were acceptable.

3.1.3. Lack of Minerals

In order to maintain normal physiological functions, wild ungulates usually search for a salt lick. Deficiency of salts can have severe negative effects on animal health, appetite, reproduction, and lactation [26]. Since captive animals are confined to limited areas in enclosures, they are more prone to mineral deficiency [27]. The current protocol included two indicators—(3.1) "availability of salt licks" and (3.2) "licking objects"—to check whether the animals were mineral-deficient. Salt-deficient individuals usually lick soil, woods, and even their own urine or that of other individuals [28]. According to the protocol for Punjab urial developed in this study, animals were scored as "mineral-deficient" if they showed any sign of licking objects or urine.

3.1.4. Thermal Ease

The natural habitat of the Punjab urial in Pakistan ranges from the dry, hot, and arid rocky mountains, to the high elevations in northern areas, which have extreme climatic conditions [29,30]. In captivity, there is a possibility of exposure for a long time to temperatures different from those of the natural habitat, which can eventually result in heatstroke and stress in animals [10]. In order to assess thermal ease, two indicators were developed: "shelter availability" (4.1) and "shade availability" (4.2). These indicators were assessed by checking whether all animals could simultaneously access shade during hot weather and shelter when climatic conditions are not favorable. These indicators were already validated in [10,19] for dorcas gazelles and domestic sheep.

3.1.5. Easiness in Movement

Wild ungulates usually roam over vast areas, and their home ranges may exceed several square kilometers [31]. Wild animals kept in relatively small enclosures are more prone to develop different negative behavioral and physiological changes. All these negative developments are clear indicators of poor animal welfare [32]. According to [33], each Punjab urial should have a minimum area of 46.45 m^2, and for every additional animal, this area should be increased by 25% (11.6 m^2). Two indicators were developed in the current protocol, which had already been used and validated in [10,11,19]. The indicators included "total area of enclosure" (5.1), and "space (m^2) offered per animal" (5.2), with the latter derived as the total enclosure area divided by the total number of animals in the area.

3.1.6. Standard Enclosures

The enclosures designed and constructed for breeding and conservation of wild animals should fulfill maximum criteria for standard enclosures. These criteria include fence structure and fixation to prevent the entry of other animals, especially small mammals. The fence should be covered so that animals do not have any visual contact with the outside environment if the aim is reintroduction. Each enclosure must have quarantine facilities, so that newly arrived animals can be held in isolation for observation, and sick or infected animals can be separated from other animals. Keeping in view the recommendations made in [34,35], four indicators, i.e., "fence structure" (6.1), "fence substratum" (6.2), "availability of quarantine" (6.3), and "number of quarantines" (6.4), were developed to assess enclosure standards.

3.1.7. Lack of Injuries

Due to limited space and the presence of aggressive animals, especially when different species are kept together, there is a chance of injuries [36]. Hoof injuries are common in artiodactylids [37]. There is a high chance of integument deformities and skin damage in captive animals. These deformities can result from any infectious disease, physical environment, aggression, or improper capture and handling of animals [10]. Two indicators were developed to assess lack of injuries: "integument deformities" (7.1) and "lameness" (7.2).

Integument deformities were assessed from close observation of the skin of animals. Hairless or damaged patches of skin of more than 2 cm were counted [10]. Animals were also observed for intraspecific aggression marks and injury marks from capture. Animals with "integument deformities" (7.1) were scored as follows: animals having several small lesions or a single lesion bigger than 2 cm^2 and hairless patches were considered to have "severe integument deformities"; animals having no lesions, but having hairless patches were considered to have "moderate integument deformities"; and animals having no lesions or hairless patches were considered to be normal with "no integument deformities".

All animals were observed for "lameness" (7.2) while they were moving. Animals having an apparent abnormality in walking or running were scored as "lame", while those having no apparent abnormality were scored as "not lame". The above indicators were already validated in [10,11,19].

3.1.8. Lack of Disease

Common diseases found in Punjab urial (*Ovis vignei punjabiensis*) are viral infections, such as contagious ecthyma, and foot and mouth diseases (FMD). The major bacterial disease is pneumonia caused by *Pasteurella* spp. [17]. Other disorders that can affect the captive animals are behavioral disorders, birth problems, gastrointestinal infections, and trauma. Trauma mostly occurs due to intraspecific aggression, during capture and handling, or due to accident [10]. According to the American Association of Zoo Veterinarians [38], every facility holding wild animals, especially for conservation and breeding purposes, must have qualified veterinarians, a veterinary facility where sick animals can be treated, and a facility fornecropsy.

Six indicators were developed to assess the lack of diseases, including respiratory and gastrointestinal diseases, and health facilities. These indicators included "ophthalmic discharge" (8.1), "nasal discharge" (8.2), "labored breathing" (8.3), "diarrhea" (8.4), "availability of veterinarian" (8.5), and "availability of veterinary facility" (8.6). If any indicators among these were observed, animals were scored as "obvious", and animals having no sign of these indicators were scored as "non-obvious". All these indicators were validated in [25,27,39,40] for the welfare assessment of cattle.

3.1.9. Displaying Social Behavior

Agonistic behavior is displayed by animals usually in response to stress, intraspecific and interspecific competition for resources, and mating competition [41]. This behavior can result in severe physical damage. *Ovis vignei punjabiensis* is a gregarious ungulate with multiple agonistic and affinitive behaviors [22]. In captivity, no studies have been conducted on these animals to evaluate their social behavior. Extreme and repeated agonistic behavior displayed by animals in captivity can be a sign of poor management and welfare [10]. On the contrary, frequently occurring affinitive behavior represents satisfaction of animals concerning their environment, and hence good animal welfare.

Two indicators were developed to assess the social behavior: "affinitive interactions" (9.1) and "agonistic interactions" (9.2). These indicators were also validated in [10,21,25]. An ethogram (Table 2) was developed based on the social behavior of Punjab urial [22], and information was obtained from a short reconnaissance survey for this study.

Table 2. List of social behaviors included in the welfare protocol for Punjab urial.

Behavior Pattern	Description of Behavior
Mutual grooming	When an animal brushes another animal with its muzzle on any part of the body with exception to the anal region. If the actor animal stops brushing for 10 s and starts again, it is to be counted as a new bout, regardless of whether the actor brushes the same receiver or another. If the actor receives reversal brushing from the receiver, it should also be counted as a new bout (AFI).
Licking	One animal licks any part of another animal with the tongue with the exception of anal region or urine. If the actor animal stops licking for 10 s and starts again, it is to be counted as a new bout, regardless of whether the actor licks the same receiver or another. If the actor receives reversal brushing from the receiver, it should also be counted as a new bout (AFI).
Smelling	One animal smells any part of another animal with the exception of the anal region or urine (Flehmen response). If the actor animal stops smelling for 10 s and starts again, it is to be counted as a new bout, regardless of whether the actor smells the same receiver or another. If an actor receives reversal smelling from the receiver, it should also be counted as a new bout (AFI).
Play	Physical contact of two animals by rubbing bodies, horning, head to head play, or rubbing horns against the neck or other parts of the body, with no signs of aggression or taking advantage. If the actor stops for more than 10 s and then resumes with the same receiver or another, it should be counted as a new bout (AFI).
Chase	One animal running behind another animal (receiver), causing the receiver to flee from its previous position. The animal also shows strong aggression toward the receiver (AGI).
Block	One animal (actor) runs after another (receiver), stands broadly in front of the receiver, and prevents approaching the opposite sex (AGI).
Parallel walk	Two animals at the same time walking parallel with heads bent down and maintaining a distance of 10–20 m. If the animals scratch the ground with their feet, stop for 10 s or more, and resume, it is to be counted as a new bout (AGI).
Fighting and thrusting	Two animals raise their front legs and strike their heads and horns, or push one another back (head-to-head or horns' base) planting legs on the ground with great force. One animal hits others with a kick or strong butting, or if any animal thrusts vegetation or another object with signs of aggression. If any animal displays such behavior and stops for 10 s or more and then resumes, it should be counted as a new bout (AGI).

AFI = Affinitive interactions, AGI = Agonistic interactions.

3.1.10. Group Dynamics

Punjab urial are gregarious ungulates [30] with different types of herds. Rams usually prefer to join herds in the rutting season; thus, the herd composition changes with the season. Three types of herds have been observed in wild Punjab urial: female herds exclusively comprising ewes and young; mixed herds containing one or more rams, ewes, and young; and male herds containing only males [22]. Husbandry guidelines suggest that in reproductive groups of captive ungulates, one adult male should be kept with 3–7 females and their young, while 3–7 adult males should be kept in bachelor groups [10]. In order to prevent inbreeding, reproductive males should be replaced or exchanged after every two years and females after three years [34].

Following [10], two indicators—"herd size" (10.1) and "herd composition" (10.2)—were developed to assess the group dynamics. Wildlife enclosures and zoos have a current tendency to build larger and more naturalistic facilities. Sometimes it is encouraged to mix species with other sympatric species where they share the same space and resources, which may serve as a source of enrichment, especially for animals raised for conservation and reintroduction purposes [41]. The current protocol included another indicator "number of animals (other species)" (10.3) [10], to record the number of individuals of other species. In addition to the number, any information about interspecific interactions based on the behavior should be recorded, which could be very useful for establishing management guidelines.

3.1.11. Display of Other Behavior

Behavior that is invariable and repetitive without any apparent purpose is termed stereotypic behavior. Most of this behavior is due to stress, frustration, repeated attempts to cope, or failures of the central nervous system [27,42]. Sometime this behavior may help animals cope with hostile and difficult environments, but stereotypic behavior is generally regarded as an indicator of poor welfare [43]. There is no such information about stereotypic behavior in captive Punjab urial. Wild animals, especially ungulates in captivity, are more prone to develop oral stereotypies [44]. This protocol suggested that all animals should be checked for the presence or absence of "stereotypic behavior" (11.1). Animals should be scored as "stereotypes present" and "stereotypies absent" in the case of presence or absence of stereotypic behavior, respectively. Any behavior that is observed and recorded should be described properly [45]. The use of this indicator was strongly recommended in the welfare assessment captive animals and was validated successfully by [10].

Captive breeding programs can be promising for species that are raised for reintroduction, if environmental enrichment programs are encouraged. *Ovis vignei punjabiensis* prefers to live in rocky mountain areas with shrubs and grasses [29]. Although there are no husbandry guidelines for this species, guidelines for other species recommend providing opportunities for animals to perform their natural behaviors of browsing, grazing, running away, and hiding. In order to fulfill these requirements, it is recommended to make use of structural components like rocks, vegetation, and uneven ground, and to modify feeding techniques periodically with accompanying simulations of sounds and smell [10]. These practices will help produce captive-born animals suitable for release into the wild for reintroduction purposes.

To assess "environmental enrichment programs" (11.2) [10], the animals' access to opportunities such as browsing and grazing should be recorded. In addition, it should be taken into account whether certain objects, such as poles or sticks of different shapes and sizes, are used; if yes, then when and how they are used must be described.

3.1.12. Good Human–Animal Affiliations

For keeping captive populations healthy, it is recommended to minimize stress usually associated with veterinary procedures [37]. Arranging proper medical training programs under the supervision of experienced staff can help to reduce stress and injuries while handling and treating animals. In case no "medical training programs" (12.1) are arranged, then it is of utmost importance that the facility must hold advanced "capturing, handling, immobilization, and translocation" (12.2) systems that can ensure the minimization of stress, damage, or any loss of animals. These indicators were already validated for dorcas gazelles in [10].

3.2. *Application of Punjab Urial Welfare Assessment Protocol*

The following results were obtained after implementation of the proposed protocol.

3.2.1. Lack of Prolonged Appetite

According to the results obtained for body conditions (1.1) after application of the protocol, all animals had normal body condition with the exception of one adult male at CU1 with excessive body condition. According to the life history record of this male, it was never observed to have mated during the rutting season.

3.2.2. Lack of Prolonged Thirst

The assessment of water sources (2.1) was conducted for all the three groups. At CU1, a concrete pond (0.91 × 0.91 m with a depth of 0.45 m) was present; CU2 had a concrete pond (1.52 × 1.52 m with a depth of 0.45 m) plus additional clay buckets (0.609 × 0.609 m); and CU3 had a concrete pond (2.74 × 2.74 m with a depth of 0.45 m) and three additional clay buckets (0.609 × 0.609 m). Water availability (2.2) and cleanliness of water sources (2.3) needed to be improved in CU2 and CU3. Both facilities had concrete ponds, but they were dry and the water in the clay buckets was not fresh and clean. Only the pond at CU1 had a sufficient supply of fresh water from a nearby spring. The study was conducted in the hot month of August and animals were frequently observed drinking water from this pond.

3.2.3. Lack of Minerals

Salt licks were available (3.1) in sufficient quantity in each of the three facilities, and no animal was found to be licking objects (3.2). According to these results, animals showed no signs of mineral deficiency.

3.2.4. Thermal Ease

Each of the three facilities provided shelters (4.1) for animals. At CU1, there were three shelters in total: two of 3.04 × 13.04 m (metal) and one of 3.96 × 3.04 m with a wooden roof. All animals were provided with enough shade (4.2) during the hot summer. In addition to the shelters, CU1 had many large sheesham (*Dalbergia sissoo*) trees providing thick cover for the animals. Animals were observed preferring to stay in the shade of trees rather than in artificial shelters. One shelter (15.24 × 6.09 m) was provided at CU2, and two shelters (15.24 × 6.09 m each) were provided at CU3. In addition, enclosures held enough vegetation (*Acacia modesta*) to provide shade for the animals. Every animal had access to shade and shelter in bad weather conditions (hot weather and rainy days).

3.2.5. Easiness in Movement

The total area of the enclosures (5.1) and the space offered per animal (5.2) varied across the enclosures. Being constructed in or on the boundaries of wildlife parks, every enclosure encompassed a large area, with sufficient natural vegetation and uneven ground. CU1 was the only enclosure which lay close to local settlements and roads (Table 3).

Table 3. Total enclosure area and area offered per animal in each herd, including individuals from other species (having the same approximate space requirements as Punjab urial).

Facility	Number of Animals	Total Area of Enclosure (Area) (m^2)	Space Per Animal (m^2/Animal)
CU1 (Cherat Wildlife Park)	23	14,299	621
CU2 (Manglot Wildlife Park)	6	3530	588
CU3 (Manglot Wildlife Park)	22 [a]	15,793	717

[a] Punjab urial, $n = 8$; chinkara, $n = 13$; mouflon sheep, $n = 3$.

Although enough area was provided, sub enclosures should be designed and constructed within the main enclosures to separate different reproductive groups.

3.2.6. Standard Enclosures

All three enclosures were assessed for existing fence conditions (6.1): whether it was covered or not, or broken. The fences in all the three enclosures were single layered and uncovered, but not broken at any point. The location of the enclosures and thick vegetation prevented the animals from eye contact with the public. In CU1, due to comparatively less vegetation and nearby road and settlements, the fence should be double wired and covered in order to avoid any physical or frequent eye contact with the public.

Fence substrata (6.2) were assessed and found to be in good condition, thus preventing the entry of other small mammals into the enclosures. None of the three enclosures had quarantine facilities (6.3 and 6.4). For robust captive breeding and management, the availability of quarantine facilities should be ensured.

3.2.7. Lack of Injuries

Animals were observed visually (also included the use of binoculars) for the assessment of integument deformities (7.1). No animal was found with any integument deformity. All animals were carefully assessed for lameness (7.2), but all the animals were normal.

3.2.8. Lack of Diseases

Eight individuals (one adult male, three adult females, and four juveniles) at CU1 were observed to have ophthalmic discharge (8.1). None of the animals from CU1, CU2, or CU3 were observed to have signs of nasal discharge (8.2), labored breathing (8.3), or diarrhea (8.4). No classified veterinarian (8.5) or veterinary facility (8.6) was present at any of the three facilities. Veterinary records of each facility revealed that animal health checks were performed once a month, and the animals were checked by a classified veterinarian in case of mass infections. Keeping in view the budget constraints, optimum veterinary facilities in all the centers should be improved accordingly.

3.2.9. Displaying Social Behavior

Results for affinitive interactions (9.1) and agonistic interactions (9.2) showed that the highest aggression was shown at CU3 (66.59%), followed by CU1 (43.64%) and CU2 (37.75%) (Figure 1). The results obtained from the Shapiro–Wilk test (W = 0.634, $p < 0.05$) were significant. Thus, our data was not normally distributed. The Kruskal–Wallis H test showed that there was a statistically significant difference in behavior among the three groups (χ^2 (2) = 10.073, $p < 0.05$), with a mean rank of 87.43 for CU1, 80.22 for CU2, and 109.00 for CU3. Animals at CU3 shared the space with other animals (Table 3).

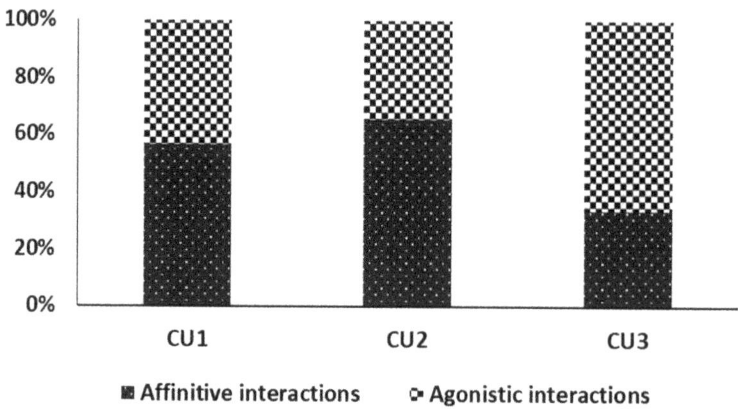

Figure 1. Percentage of affinitive and agonistic interactions of the total number of interactions recorded for each herd of Punjab urial assessed (n = 37).

While assessing the social behavior (9.1 and 9.2), the Punjab urial at CU3 showed aggression (chasing and head jerking) toward the chinkara. No obvious aggression was observed from Punjab urial toward mouflon sheep. All three species were observed maintaining approximately a distance of 10–20 m from each other.

3.2.10. Group Dynamics

An assessment of the number of Punjab urial (10.1) in all the centers showed that only animals at CU3 were sharing space with other species, including chinkara (*Gazella bennettii*; n = 13) and mouflon sheep (*Ovis orientalis*; n = 3) (10.3). Results for assessing herd composition (10.3) were: CU1, n = 23 animals (adult male = 5, subadult male = 1, female = 6, young = 11); CU2, n = 6 (adult male = 3, adult female = 1, young = 2); and CU3, n = 8 (adult male = 4, adult female = 2, young = 2). Life history records of Punjab urial at each of the three facilities showed that no male had been replaced with other males from a different population. This situation was alarming and could potentially lead to inbreeding depression and drastic downfall of the population. These results strongly suggested that the exchange of productive males and females between different populations is necessary to avoid inbreeding.

3.2.11. Display of Other Behavior

While assessing animals for stereotypic behavior (11.1), only two adult females and one sub-adult male at CU1 were found to move back and forth repetitively next to a fence and bite the fence. Regarding environmental enrichment (11.2), all the centers provided natural opportunities to forage, in the form of thick and tall vegetation for browsing. Due to comparatively fewer shrubs and less low vegetation at CU1, formal enrichment in the form of vertical poles or sticks to offer food should be encouraged. This will increase complexity in the physical environment of the enclosures.

3.2.12. Good Human–Animal Affiliations

Each of the herds assessed at the three different centers had never been a part of any medical program (12.1). Very poor results were recorded regarding capturing, handling, immobilization, and translocation techniques in all the three centers. Animal capture was mostly conducted by chasing animals and confining them to a corner of the enclosure. Veterinary and post-mortem reports revealed the expiry of seven chinkaras and three Punjab urials during capturing and translocation, respectively.

3.3. Overall Comparison

Overall, all the three herds were same in terms of resource-based indicators. A slight difference among the three herds was present in terms of animal-based indicators. The herd at CU1 (43.64%) showed aggression and some animals (34.78%) were observed showing signs of ophthalmic discharge. The second herd showed an aggression of 37.75%, which was comparatively lower than that observed in the other two herds, but no diseased animal was found. The herd at CU3 showed the highest aggression (66.59%), but no animal was diseased. Comparing the results obtained for all the three groups, we assumed the herd at CU2 was at maximum welfare, followed by CU3 and CU1.

4. Discussion

The current project was designed to develop, for the first time in Pakistan, a baseline protocol to assess the welfare of wild ungulates in captivity. The assessment of resources, management, and animal-based measures are collectively termed as animal welfare assessments. In order to have a clear idea about the welfare of a group of animals in captivity, it is difficult to achieve the goals by having a single indicator or very few indicators. Thus, to have a complete and appropriate welfare assessment for a particular species, it is important to have a combination of several indicators according to the biology and ecology of that species [21]. The protocol proposed in this study is based on the welfare protocol for domestic sheep. This protocol differs from the domestic sheep welfare protocol in terms of the number of indicators, consisting of an extended list of 31 different animal- and resource-based indicators. In this newly developed protocol, we found that some indicators are difficult to assess with accurate results, especially in wild animals, when they are kept in large enclosures with dense vegetation. Although we used binoculars to assess integument and skin deformities (indicator 7.1), it was difficult to determine if there were any small lesions or patches on the body. According to [46], there is a high possibility of high levels of aggression and fights in wild ungulates in captivity. Detailed observations of skin and other integuments are thus very important. We do not encourage excluding this indicator; rather, we suggest the use of more powerful binoculars or a high-resolution camera to obtain clear pictures of the animals.

Our protocol is based on the welfare protocol for domestic sheep, but we excluded the criteria 'positive emotional state'. There is a lack of information relative to this subject in captive ungulates [10]. This state includes pleasure, comfort, confidence, and interest. The aim of animal welfare assessment is to determine positive emotional states, or reduce undesirable experiences and increase opportunities for animals to have more healthy and positive states [47]. We suggest that these criteria should be included in developing welfare assessment protocols for any wild species in captivity.

An emerging trend in establishing zoos and enclosures is to provide large and more natural environments for captive animals [48]. The current protocol found the area requirements for captive Punjab urial very true-to-life and acceptable in each of the three enclosures examined. All three enclosures offered vast areas to the animals, with natural habitats utilizing natural vegetation and uneven ground. All the animals could easily experience grazing, browsing, and athletic activities. With the exception of CU1, where animals were observed to be moving back and forth and the enclosure is located in close proximity to local settlements and roads, so stereotypic behavior was recorded during the application of the current protocol. According to [49], eye contact of visitors and wild animals in captivity can result in stress and stereotypic behavior. Following [34], fences at CUI should be covered with raffia in order to avoid frequent eye contact between the animals and the public.

Interspecific aggression has been mostly documented in carnivores [50], while such information for wild ungulates is scarce [51]. Interspecific aggression can possibly increase intraspecific aggression. In our study, we found that animals in CU3 showed the highest aggression (66.59%), followed by CU1 (43.64%) and CU2 (37.75%). We assumed that the higher aggression in CU3 was due to the presence of other species [52]. Punjab urial males were frequently observed chasing chinkara, showing comparatively less aggression toward mouflon sheep. We also recorded counter aggression from chinkara males toward Punjab urial and mouflon sheep. According to [36], interspecific aggression is usually greater between distantly related species than closely related species, and it is recommended to separate the species with aggressive males. Our results suggested isolating chinkara from Punjab urial. Regarding stereotypic behavior, it was observed only in three animals (two adult females and one subadult male) at CU1. This facility had less natural vegetation as compared to the other two facilities, where the animals did not show any stereotypic behavior. According to [45], captive ungulates have a high tendency to produce oral stereotypies when they have limited opportunities for natural foraging; those findings are in agreement with results produced from the current study.

Recently, medical training programs and training techniques have been practiced and understood in modern zoos and facilities. These methodologies are frequently used and applied in different species, including big cats, elephants, giraffes, and apes. There is a lack of implementation of such programs in ungulates [10]. We consider it important to add a medical training program (12.2) because, if properly practiced, it is a promising means of reducing the stress caused by veterinary techniques. For capture of animals, every facility must have the right material (capture enclosure, net, handling crush, and dart gun), coupled with an experienced team, in order to avoid trauma and other serious injuries [34]. During application of the proposed protocol, it was found that several animals had expired during capture and translocation. These results make the medical training program a top priority in Punjab urial and other associated captive ungulates.

The development of this welfare assessment protocol is a leading documented work in developing a scientific and standard tool for the measurement of welfare in Punjab urial *(Ovis vignei punjabiensis)*. Using the Welfare Quality® protocol for farm animals as a reference, welfare assessment protocols for several wild species have been developed, including those for mink (*Neovison vison*), foxes (*Vulpes* spp.) [47], dorcas gazelles (*Gazella dorcas*) [10], and bottlenose dolphins (*Tursiops truncatus*) [47,53].

Welfare assessment protocols are developed with the aim of assessing the welfare level in captive animals, to discover limitations in captive breeding, and to ensure optimal welfare through recommendations. Protocols that are practical and easy to be applied are considered successful protocols. Protocols developed for different species differ in terms of time for their implementation. In the case of mink and foxes [47], the developed welfare assessment protocol needs three visits to each farm. In the case of bottlenose dolphins [53], the protocol requires two days for a complete welfare assessment of a dolphin pod including up to 10 individuals. Our protocol includes an extensive set of 31 indicators, and thus has some practical challenges. According to [10], protocols for dorcas gazelles require less than 6 h per herd. During the application of the proposed protocol, the largest herd of Punjab urial (n = 23) was assessed at CU1, and all the indicators were assessed in 5 h per herd. Our protocol is in early-stage development for assessing the welfare in captive Punjab urial and endorsement is still needed; however, its application to the three different herds and the results obtained allowed to identify some areas in all the facilities which need to be improved.

5. Conclusions

Using the welfare assessment protocol for farm animals (sheep) as the base, we developed a welfare protocol for welfare assessment in captive Punjab urial (*Ovis vignei punjabiensis*). This first specific protocol developed for Punjab urial comprised 4 basic principles, 12 criteria, and 31 animal- and environmental-based indicators. Although this protocol still needs validation, its first application and subsequent results obtained from three different herds of captive Punjab urial at CU1, CU2, and CU3 highlight some areas where improvements are essential. According to the results obtained from this protocol, handling, capturing, and translocation were found to be the most important areas for necessary action as most of the mortalities happened in capturing and translocation. Another important area which needs to be improved on an urgent basis is the availability of veterinary facilities. Furthermore, we recommend the shifting of breeding animals between different populations. Based on the results, we recommend covering the fence at CU1 in order to prevent frequent human–animal eye contact. On the basis of this study results, we believe that this protocol can be a promising tool for welfare assessment at facilities that hold Punjab urial. The current protocol has the best combination of welfare indicators for the target species and is a leading step in captive breeding research in Pakistan. In addition, this protocol can be used as a base for developing similar welfare protocols for other captive mountain ungulates in Pakistan and globally.

Author Contributions: Conceptualization, R.H.K. and Z.L.; data curation, R.H.K.; formal analysis, R.H.K. and L.T.; funding acquisition, Z.L.; investigation, R.H.K. and L.T.; methodology, R.H.K.; project administration, R.H.K., L.T. and Z.L.; resources, R.H.K. and Z.L.; software, R.H.K. and L.T.; supervision, Z.L.; validation, L.T. and Z.L.; visualization, R.H.K. and L.T.; writing—original draft, R.H.K.; writing—review and editing, R.H.K., L.T. and Z.L.

Funding: This study was supported by the National Nature Science Foundation of China (31372221, 31870512), and the fundamental research funds for the central universities (2572014CA03, DL13EA01). This study also got financial support from Heilongjiang Touyan innovation team Program for forest ecology and conservation.

Acknowledgments: We are grateful to the Khyber Pakhtunkhwa Wildlife Department for allowing us to carry our study. We are also thankful to the staff members of every facility (CWP and MWP) for their cooperation throughout this study. We are thankful to Muhammad Zada for his valuable help. Authors are extremely grateful to Nazneen Sultan for giving valuable suggestions throughout the study.

Conflicts of Interest: The authors declare no conflicts of interest.

References

1. Salas, M.; Temple, D.; Abáigar, T.; Cuadrado, M.; Delclaux, M.; Enseñat, C.; Almagro, V.; Martínez-Nevado, E.; Quevedo, M.Á.; Carbajal, A.; et al. Aggressive behavior and hair cortisol levels in captive *Dorcas gazelles* (*Gazella dorcas*) as animal-based welfare indicators. *Zoo Biol.* **2016**, *35*, 467–473. [CrossRef] [PubMed]
2. Kleiman, D.; Allen, M.; Thompson, K.; Lumpkin, S. Wild mammals in captivity: Principles and techniques. *Biol. Conserv.* **1998**, *2*, 232.
3. Witzenberger, K.A.; Hochkirch, A. Ex situ conservation genetics: A review of molecular studies on the genetic consequences of captive breeding programmes for endangered animal species. *Biodivers. Conserv.* **2011**, *20*, 1843–1861. [CrossRef]
4. Williams, S.E.; Hoffman, E.A. Minimizing genetic adaptation in captive breeding programs: A review. *Biol. Conserv.* **2009**, *142*, 2388–2400. [CrossRef]
5. Gusset, M.; Dick, G. Building a Future for Wildlife? Evaluating the contribution of the world zoo and aquarium community to in situ conservation. *Int. Zoo Yearb.* **2010**, *44*, 183–191. [CrossRef]
6. Liz, P. *Beyond Captive Breeding: Re-Introducing Endangered Mammals to the Wild: Zoological Society of London Symposia 62*; Gipps, J.H.W., Ed.; Clarendon Press: Oxford, UK, 1991; Volume 26, p. 56. ISBN 0-19-854019-1.

7. Benefiel, A.C.; Dong, W.K.; Greenough, W.T. Mandatory "Enriched" Housing of Laboratory Animals: The Need for Evidence-based Evaluation. *ILAR J.* **2005**, *46*, 95. [CrossRef]
8. World Association of Zoos and Aquariums. Ethics and Animal Welfare. In *Building a Future for Wildlife—The World Zoo and Aquarium Conservation Strategy*; Olney, P.J.S., Ed.; WAZA Executive Office: Bern, Switzerland, 2005; pp. 59–64. ISBN 303300427X.
9. Gosling, S.D. From mice to men: What can we learn about personality from animal research? *Psychol. Bull.* **2001**, *127*, 45. [CrossRef]
10. Salas, M.; Manteca, X.; Abáigar, T.; Delclaux, M.; Enseñat, C.; Martínez-Nevado, E.; Quevedo, M.; Fernández-Bellon, H. Using farm animal welfare protocols as a base to assess the welfare of wild animals in captivity Case study: *Dorcas gazelles* (*Gazella dorcas*). *Animals* **2018**, *8*, 111. [CrossRef]
11. Welfare Quality®. *Welfare Quality® Assessment Protocol for Cattle*; Welfare Quality®Consortium: Lelystad, The Netherlands, 2009; ISBN 9789078240044.
12. Ayaz, S.M.; Jamil, A.A.; Khan, M.; Ayaz Qamar, M.F. Behaviour and biology of *Ovis orientalis* (urial) in Kotal Wildlife Park and Borraka Wildlife sanctuary in Kohat. *J. Anim. Pl. Sci.* **2012**, *22*, 29–31.
13. Damm, G.R.; Franco, N.S. *CIC Caprinae Atlas of the World*; CIC International Council for Game and Wildlife Conservation: Budakeszi, Hungary, 2014.
14. Roberts, T.J. *The Mammals of Pakistan*; Earnst Benn Limited: London, UK, 1977; pp. 206–208.
15. International Union for Conservation of Nature and Natural Resources. *International Red List of Threatened Species 2008*; International Union for Conservation of Nature and Natural Resources: Gland, Switzerland, 2008.
16. Sheikh, K.; Molur, S. *Status and Red List of Pakistan Mammals, Based on Conservation Assessment and Management Plan for Mammals*; International Union for Conservation of Nature and Natural Resources: Islamabad, Pakistan, 2005; p. 344.
17. Awan, G.A.; Ahmad, T.A.H.I.R.A.; Festa-Bianchet, M.A.R.C.O. Disease spectrum and mortality of Punjab urial (*Ovis vignei punjabiensis*) in Kalabagh Game Reserve. *Pak. J. Zool.* **2005**, *37*, 175.
18. International Union for Conservation of Nature and Natural Resources. *International Red List of Threatened Species 2002*; International Union for Conservation of Nature and Natural Resources: Gland, Switzerland, 2002.
19. Dwyer, C.; Ruiz, R.; Beltran de Heredia, I.; Canali, E.; Barbieri, S.; Zanella, A. *AWIN Welfare Assessment Protocol for Sheep*; Animal Welfare Indicators (AWIN): Edinburgh, UK, 2015. [CrossRef]
20. Mirza, Z.B. *Study of Morphology, Distribution, Population, Behaviour, Food and Habitat of Punjab Urial (Ovis orientalis punjabiensis), in Salt Range and Kala Chitta Range, Punjab, Pakistan*; Pakistan Museum of Natural History: Islamabad, Pakistan, 1980.
21. Botreau, R.; Veissier, I.; Butterworth, A.; Bracke, M.B.; Keeling, L.J. Definition of criteria for overall assessment of animal welfare. *Anim. Welf.* **2007**, *16*, 698–705.
22. Schaller, G.B.; Mirza, Z. On the behaviour of Punjab urial (Ovis orientalis punjabiensis). In *The Behaviour of Ungulates and Its Relation to Management*; The University Of Calgary: Calgary, AB, Canada, 1974; pp. 306–323.
23. Fred, K.; Kumarasinghe, J.C. Remarks on body growth and phenotypes in Asian elephant *Elephas maximus*. *Acta Theriol. Suppl.* **1998**, *5*, 135–153.
24. Audigé, L.; Wilson, P.; Morris, R. A body condition score system and its use for farmed red deer hinds. *N. Z. J. Agric. Res.* **1998**, *41*, 545–553. [CrossRef]
25. Hernandez, A.; Berg, C.; Eriksson, S.; Edstam, L.; Orihuela, A.; Leon, H.; Galina, C. The Welfare Quality® assessment protocol: How can it be adapted to family farming dual purpose cattle raised under extensive systems in tropical conditions? *Anim. Welf.* **2017**, *26*, 177–184. [CrossRef]
26. Michell, A. Sodium and research in farm animals: Problems of requirement, deficit, and excess. *Outlook Agric.* **1985**, *14*, 179–182. [CrossRef]
27. Mason, G.; Rushen, J. *Stereotypic Animal Behaviour: Fundamentals and Applications to Welfare*; Centre for Agriculture and Bioscience International: Wallingford, UK, 2006.
28. Underwood, E.J. The mineral nutrition of livestock. *Vet. J.* **1993**, *161*, 70–71.
29. Shackleton, D.M. Social maturation and productivity in bighorn sheep: Are young males incompetent? *Appl. Anim. Behav. Sci.* **1991**, *29*, 173–184. [CrossRef]
30. Roberts, T.J. *The Mammals of Pakistan*; Oxford University Press: Karachi, Pakistan, 1997; p. 299.
31. Parrini, F.; Grignolio, S.; Luccarini, S.; Bassano, B.; Apollonio, M. Spatial behaviour of adult male Alpine ibexCapra ibex ibex in the Gran Paradiso National Park, Italy. *Acta Theriol.* **2003**, *48*, 411–423. [CrossRef]

32. Carlstead, K.; Shepherdson, D. Alleviating stress in zoo animals with environmental enrichment. In *The Biology of Animal Stress: Basic Principles and Implications for Animal Welfare*; Centre for Agriculture and Bioscience International: Wallingford, UK, 2000; pp. 337–354.
33. America, Z.A. Animal Care & Enclosure Standards and Related Policies. 2016. Available online: http://www.zaa.org (accessed on 12 September 2019).
34. Espinosa, J.; López-Olvera, J.R.; Cano-Manuel, F.J.; Fandos, P.; Pérez, J.M.; López-Graells, C.; Ráez-Bravo, A.; Mentaberre, G.; Romero, D.; Soriguer, R.C.; et al. Guidelines for managing captive Iberian ibex herds for conservation purposes. *J. Nat. Conserv.* **2017**, *40*, 24–32. [CrossRef]
35. Jordan, B. Science-based assessment of animal welfare: Wild and captive animals. *Rev. Sci. Tech. Off. Int. Epizoot.* **2005**, *24*, 515. [CrossRef]
36. Popp, J.W. Interspecific aggression in mixed ungulate species exhibits. *Zoo Biol.* **1984**, *3*, 211–219. [CrossRef]
37. Boever, W. Artiodactylids (Artiodactyla): Noninfectious diseases. In *Zoo and Wild Animal Medicine*, 2nd ed.; Fowler, M.E., Ed.; WB Saunders & Co.: Philadelphia, PA, USA, 1986; pp. 962–964. ISBN 0721610137.
38. Carpenter, N.S.C.; Helmick, K.; Meehan, T.; Murray, M.; Smith, J.; Wyatt, J. *Guidelines for Zoo and Aquarium Veterinary Medical Programs and Veterinary Hospital*, 6th ed.; American Association of Zoo Veterinarians: Yulee, FL, USA, 2016.
39. Kaurivi, Y.; Laven, R.; Hickson, R.; Stafford, K.; Parkinson, T. Identification of Suitable Animal Welfare Assessment Measures for Extensive Beef Systems in New Zealand. *Agriculture* **2019**, *9*, 66. [CrossRef]
40. Battini, M.; Vieira, A.; Barbieri, S.; Ajuda, I.; Stilwell, G.; Mattiello, S. Invited review: Animal-based indicators for on-farm welfare assessment for dairy goats. *J. Dairy Sci.* **2014**, *97*, 6625–6648. [CrossRef] [PubMed]
41. Lovari, S.; Fattorini, N.; Boesi, R.; Bocci, A. Male ruff colour as a rank signal in a monomorphic-horned mammal: Behavioural correlates. *Sci. Nat.* **2015**, *102*, 39. [CrossRef] [PubMed]
42. Mason, G.J. Stereotypies: A critical review. *Anim. Behav.* **1991**, *41*, 1015–1037. [CrossRef]
43. Loijens, L.W.S.; Schouten, W.G.P.; Wiepkema, P.R.; Wiegant, V.M. Brain Opioid Receptor Density Relates to Stereotypies in Chronically Stressed Pigs. Stress. *Int. J. Biol. Stress* **1999**, *3*, 17–26. [CrossRef]
44. Manteca, X.; Salas, M.; Temple, D. Animal-based indicators to assess welfare in zoo animals. *CAB Rev.* **2016**, *11*, 1–10. [CrossRef]
45. Manteca, X.; Salas, M. *Concepto de Bienestar Animal*; Zawec Zoo Animal Welfare Education Center: Barcelona, Spain, 2015.
46. Abáigar Ancín, T.; López Jiménez de Rueda, L. *Husbandry Guidelines for Captive Breeding and Management of Saharawi Dorcas gazelle (Gazella dorcas neglecta)*; Spanish National Research Council: Madrid, Spain, 2013; pp. 1–345.
47. Mononen, J.; Møller, S.H.; Hansen, S.W.; Hovland, A.L.; Koistinen, T.; Lidfors, L.; Malmkvist, J.; Vinke, C.M.; Ahola, L. The Development of on-farm welfare assessment protocols for foxes and mink: The WelFur project. *Anim. Welf. UFAW J.* **2012**, *21*, 363. [CrossRef]
48. Reade, L.S.; Waran, N.K. The modern zoo: How do people perceive zoo animals? *Appl. Anim. Behav. Sci.* **1996**, *47*, 109–118. [CrossRef]
49. Sherwen, S.L.; Hemsworth, P.H. The visitor effect on zoo animals: Implications and opportunities for zoo animal welfare. *Animals* **2019**, *9*, 366. [CrossRef]
50. Palomares, F.; Caro, T.M. Interspecific killing among mammalian carnivores. *Am. Nat.* **1999**, *153*, 492–508. [CrossRef] [PubMed]
51. Ferretti, F.; Sforzi, A.; Lovari, S. Behavioural interference between ungulate species: Roe are not on velvet with fallow deer. *Behav. Ecol. Sociobiol.* **2010**, *65*, 875–887. [CrossRef]
52. Fattorini, N.; Brunetti, C.; Baruzzi, C.; Macchi, E.; Pagliarella, M.C.; Pallari, N.; Lovari, S.; Ferretti, F. Being "hangry": Food depletion and its cascading effects on social behaviour. *Biol. J. Linn. Soc.* **2018**, *125*, 640–656. [CrossRef]
53. Clegg, I.; Borgerturner, J.; Eskelinen, H. C-Well: The development of a welfare assessment index for captive bottlenose dolphins (*Tursiops truncatus*). *Anim. Welf.* **2015**, *24*, 267–282. [CrossRef]

Article

Using Thermal Imaging to Monitor Body Temperature of Koalas (*Phascolarctos cinereus*) in A Zoo Setting

Edward Narayan [1,2,*], Annabella Perakis [2] and Will Meikle [3]

1 School of Agriculture and Food Sciences, Faculty of Science, University of Queensland, QLD 4072, Australia
2 School of Science and Health, Western Sydney University, Penrith NSW 2751, Australia; 18024959@student.westernsydney.edu.au
3 Wildlife Sydney Zoo, 1-5 Wheat Rd, Sydney NSW 2000, Australia; Will.Meikle@merlinentertainments.com.au
* Correspondence: e.narayan@uq.edu.au

Received: 17 November 2019; Accepted: 5 December 2019; Published: 6 December 2019

Simple Summary: Body temperature regulation is integral for the health and well-being of animals, especially in Zoo settings. Endothermic vertebrates such as small mammals are able to maintain a constant internal body temperature; however, extreme variation in body temperature may be reflective of underlying injuries or health issues. Thus, new technology that can enable the measurement of body temperature of small mammals without need for capture and handling can be very useful for the monitoring of animals in Zoos. In this study, we report the application of an IR thermal imaging camera for monitoring the body temperature of koalas. We found that the eye and abdomen were the most consistent body features to record body temperature. This tool will have useful application for welfare evaluation of small mammals, such as koalas in Zoos.

Abstract: Non-invasive techniques can be applied for monitoring the physiology and behaviour of wildlife in Zoos to improve management and welfare. Thermal imaging technology has been used as a non-invasive technique to measure the body temperature of various domesticated and wildlife species. In this study, we evaluated the application of thermal imaging to measure the body temperature of koalas (*Phascolarctos cinereus*) in a Zoo environment. The aim of the study was to determine the body feature most suitable for recording a koala's body temperature (using coefficient of variation scores). We used a FLIR530^{TM} IR thermal imaging camera to take images of each individual koala across three days in autumn 2018 at the Wildlife Sydney Zoo, Australia. Our results demonstrated that koalas had more than one reliable body feature for recording body temperature using the thermal imaging tool—the most reliable features were eyes and abdomen. This study provides first reported application of thermal imaging on an Australian native species in a Zoo and demonstrates its potential applicability as a humane/non-invasive technique for assessing the body temperature as an index of stress.

Keywords: thermal imaging; koalas; body temperature; heat/cold stress; thermoregulation; substrate; welfare; Zoo

1. Introduction

Endothermic animals use various physiological and behavioural mechanisms to control body heat production and loss so that their internal body temperature remains constant, and this is considered "thermoregulation" [1]. The ability for an individual to thermoregulate allows them to cope with a range of environmental temperatures (i.e., extreme heat or cold), which could otherwise have detrimental effects on an animal's well-being [2,3]. Specifically, "behavioural" thermoregulation depends on the spatial arrangement and availability of microclimates in an individual's environment; most species can exploit cooler environments during hot weather and warmer environments during cold weather [4]. A suitable thermal landscape in wildlife enclosures is essential for captive management and animal welfare [5,6]. Zoo enclosures may have heating and cooling systems to provide suitable surrounding temperatures for animals [7], however it is often difficult to monitor body temperature of animals without handling. Recent technology advancements have allowed for thermal imaging to be a viable technique for measuring body temperature from a distance. An advantage of using thermal imaging is its high resolution, ability to contrast variation in body temperature, and non-invasive nature [8–10]. Thermal imaging has been used in veterinary medicine to detect leg and hoof problems using body temperature variation in racehorses, demonstrate high body temperature in livestock during transportation, and measure rapid changes in skin temperature in response to acute stressors [11,12]. To our knowledge, there is no published report on thermal imaging of native Australian wildlife in a Zoo setting.

Koalas are an arboreal folivore native to Australia [13] that use thermoregulation as a key survival mechanism to cope with subtle environmental changes, such as heat waves [3,4]. Koalas have an average body temperature between 35.5 °C and 36.8 °C and "behavioural" thermoregulation is important for maintenance of core body temperature [4,5]. Such behavioural thermoregulation in koalas includes panting to cool themselves, curling up into a ball to trap internal heat when cold, and the use of tree trunks as cold or warm substrates to regulate body temperature dependent on environmental conditions [4,5]. The selection of a perch by koalas has been demonstrated to be influenced by temperature, food availability, and time of day [4]. Interestingly, in response to temperature and "behavioural" thermoregulation, tree structure is an important consideration for koalas when selecting a perch, irrespective of food preference, because dense tree canopies provide a cool microclimate during hot weather [3–5]. Therefore, in koalas alone it is apparent that unsuitable microclimates and perch substrates in captivity can create difficulties to mimic natural thermoregulation and pose as potential stressors [14].

This preliminary study sought to record the body temperature of koalas using a thermal imaging camera to assess its applicability in a Zoo setting and specifically determine the most reliable body feature for taking body temperature readings. The application of this technology will help zookeepers to better understand the thermal preferences of koalas in relation to captive environments.

2. Materials and Methods

2.1. Study Population

Three captive-born and raised koalas at the Wildlife Sydney Zoo were observed: a 13-year-old female that has had multiple joeys (Erica), a 15-month-old juvenile male offspring of Erica (Birri), and a 16-month-old juvenile male (Alfie). The female koalas were socially housed in a large (62 m^2) open enclosure consisting of seven tree branches. Juvenile male koalas are also socially housed (three koalas), but separately to the females, in smaller less exposed enclosures (23.4 m^2) consisting of four branch perches.

2.2. Data Collection

All thermal images were taken using the FLIR530™ Thermal Imaging Camera. Thermal images of each of the koalas were taken every 30 min between 8:00 and 17:00 (9 h a day) over three consecutive days in May 2018 for a total of 171 images (57 images per individual). Each image taken was focused and centred to the koala to incorporate as much of the body features within the image (Figure 1). Each image was taken within 1 m distance to the koala.

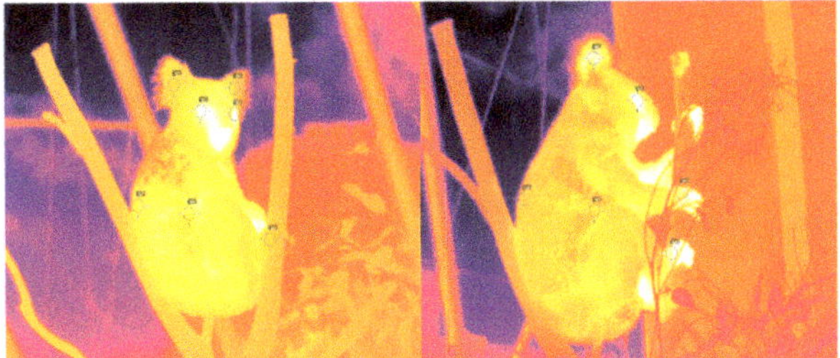

Figure 1. Thermal images of Erica on 14 (**left**) and 16 (**right**) May 2018 at Wildlife Sydney Zoo taken using a FLIR530 Thermal Imaging Camera. Both images show a minimally obstructed view of Erica perched on a branch, with the specifically selected body features highlighted by the black circular reticles.

2.3. *Image Analysis*

All images were analysed using FLIR Tools version 5.13 software for extraction of thermal metrics. The FLIR Tools software enabled us to measure the temperature of specifically selected body features of a koala by simply defining them within each image (Figure 1). We were most interested in measuring the temperature of the eyes, ears, paws, abdomen, and back.

The coefficient of variation (the ratio of the standard deviation to the mean) was calculated for each body feature (eyes, ears, paws, abdomen, or back) to determine which was the most consistent for measuring body temperature using a thermal imaging camera, i.e., the body features with the least amount of variation.

3. Results

The eyes were the most consistent and hottest body feature for measuring the body temperature of koalas by a substantial margin—coefficient of variation: 4.83%, and average temperature: 31.16 °C (Figure 2). However, the eyes were one of the least available body features to measure being visible in only 40.35% of the images. The abdomen was the second most consistent body feature with a coefficient of variation of 11.01% (Figure 2) and was visible in 48.54% of the images. The back was the most available body feature to measure (69.60% of images) and was the third most consistent to measure body temperature (13.73%; Figure 2). The paws were the least available body feature to measure (39.18% of images) and were the fourth most consistent body feature (18.12%; Figure 2). The least consistent body feature to measure body temperature was the ears (19.68%; Figure 2), which were visible in 61.99% of images. For further details on the specific eyes, ears, or paws, refer to Table 1.

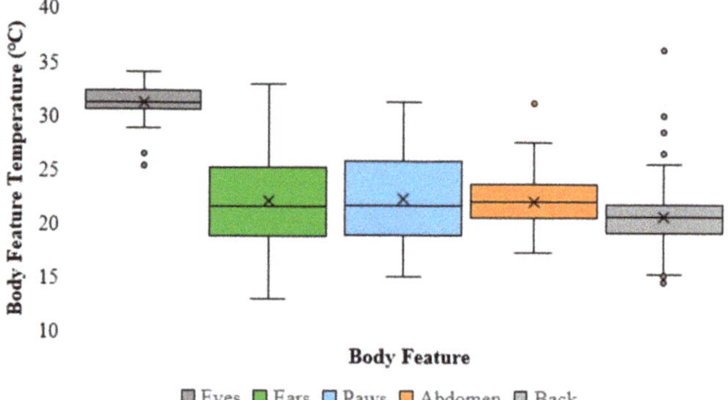

Figure 2. Boxplot demonstrating the range/spread of temperature (°C) for each body feature of a koala measured (from left to right: eyes; ears; paws; abdomen; back). The median (the line through each coloured box), average ("x"), and outliers (coloured circles) are shown for each body feature.

Table 1. Table displaying the sample size, average temperature (°C), and coefficient of variation for each of the five primary body features and the specific features that are entailed.

	Sample Size	Average Temperature (°C)	Coefficient of Variation (%)
Right Eye	45	31.27	4.36
Left Eye	42	30.63	8.65
Eyes	69	31.16	4.83
Right Ear	72	21.74	19.56
Left Ear	74	21.64	21.27
Ears	106	22.11	19.68
Right Front Paw	26	23.14	18.48
Left Front Paw	23	22.27	15.00
Right Back Paw	25	21.66	20.12
Left Back Paw	27	21.66	19.78
Front paws	43	22.55	16.25
Back Paws	47	21.48	20.07
Paws	67	22.36	18.12
Abdomen	83	22.13	11.01
Back	119	20.74	13.73

4. Discussion

Thermoregulation is the control of heat production and heat loss of an animal so that the internal body temperature remains at a more or less constant temperature; it is a property that most mammals such as koalas posses [12]. Specifically, koalas are endothermic marsupials, which have a thermal neutral zone (i.e., temperature tolerance) between 20 °C and 40 °C [1–5]; however, their average body temperature is regulated between 35.5 °C and 36.8 °C [1,3,4]. Our results from measuring these body temperatures using a thermal imaging camera (FLIR530^{TM}) suggested that the eye was the most reliable external body feature for a koala. Using these measurements of body temperature and that of perch substrate we found that koalas use thermoregulation and thermal preference of substrates on which they were perched to regulate their body temperature (Figures 1 and 2).

We were able to demonstrate in koalas that the eye was the most reliable external body feature for measuring body temperature. The lacrimal caruncle is small fleshy nodule located in the corner of the inner eye and is the hottest point of the eye, and thus the most representative external feature of core body temperature [10,15,16]. It is likely that this feature was reponsible for yielding eye temperature as our most accurate and representative reading of body temperature and that of previous studies [10,15]. Previously other studies have successfully used thermal imaging to measure core body temperature, including studies by [17], which confirms the use of eye temperature on unrestrained wild blue tits (*Cyanistes caeruleus*) as a non-invasive measure of physiological stress. Stewart et al. [18] further validates the uses of eye temperature by correlating core temperature in cattle as an indication for stress and pain. However, we could not confirm any correlation between our readings and core body temperature due to having no benchmark readings of core body temperature as measured by a tested and more invasive technique, such as a intraperitoneal device or rectal temperature reading [1]. Previous similar research [1] that did use a benchmark body temperature reading (intraperitoneal device) for comparison with various techniques demonstrated no correlation between thermal images and core body temperature in koalas. Nevertheless, [1] only used thermal images of the foot, a body feature which we have established as one of the least reliable, and thus future work investigating the correlation between thermal image readings of the eye and core body temperature in koalas could provide valuable insight. Johnson et al. [9] reported that from the thermal camera alone they were unable to give an accurate predication of core body temperature and were only able to give a correlation of it. We were able to validate the use of the eyes by not only measuring its temperature but that of other important body features for comparison of variation and repeatability. Of the three koalas measured for body temperature, all demonstrated a positive relationship between body temperature and perch substrate temperature, except, only Birri and Erica had a significant relationship.

Study Limitations

Being able to identify a reliable body feature to measure the body temperature of an animal with a thermal camera introduces a non-invasive method for measuring animal welfare. Focusing on one body part reduces outliers and abnormally high readings (Birri's max temp = 46.5 °C) that are also picked up by the thermal camera. The camera reads reflections and shadows that are present in the images therefore the camera is not reading the true temperature of the koala. This limitation has been found in similar studies, such as [10], where the solar radiation produced erratic temperature readings when reflected off other objects in the enclosure. This can been seen in the readings of the back and feet as there is such a high variation between the temperatures during the day for some koalas compared to another koalas in the same area. This variance is reflected in the CV%. The back and feet have the least consistent readings as koalas tend to sit in positions on the side of the trees or in branches which allows the sun to reflect off their backs. The nose of the koala was difficult to image because of the variability in positioning of the koala on the perches, hence it was not used in the image analysis. The camera was also picking up reflections off the artificial tree in the enclosure. Other limitations that are expressed in this study are distance from the thermal camera to the subject, as well as plants and other objects obstructing the analysed object.

Author Contributions: Conceptualization, E.N., W.M., A.P.; methodology, E.N., W.M., A.P.; formal analysis, E.N., A.P.; investigation, E.N., A.P., W.M.; resources, E.N., W.M.; writing—original draft preparation, A.P.; writing—review and editing, E.N.; project administration, E.N.

Funding: This research received no external funding.

Acknowledgments: Thank you to James Vandersteen who carried out the data analysis under Edward Narayan's supervision. We thank the staff at the Wildlife Sydney Zoo for their valuable support and providing access to the study site.

Conflicts of Interest: The authors declare no conflict of interest.

References

1. Adam, D.; Beard, L.; Johnston, S.D.; Nicolson, V.; Lisle, A.; McKinnon, A.; Larkin, R.; Theilemann, P.; Gillett, A.; Brackin, K.; et al. Recording body temperature in koalas (*Phascolarctos cinereus*): A comparison of techniques. *Aust. Vet. J.* **2018**, *96*, 308–311. [CrossRef] [PubMed]
2. Dillion, M.E.; Wang, G.; Garrity, P.A.; Huey, R.B. Review: Thermal preference in Drosophila. *J. Therm. Biol.* **2009**, *34*, 109–119. [CrossRef]
3. Crowther, M.S.; Lunney, D.; Lemon, J.; Stalenberg, E.; Wheeler, R.; Madani, G.; Ross, K.A.; Ellis, M. Climate-meditated habitat selection in an aboreal folivore. *Ecography* **2014**, *37*, 336–343.

4. Briscoe, N.J.; Handasyde, K.A.; Griffiths, S.R.; Porter, W.P.; Krockenberger, A.; Kearney, M.R. Tree-hugging koalas demonstrate a novel thermoregulatory mechanism for arboreal mammals. *Biol. Lett.* **2014**, *10*, 1–5. [CrossRef] [PubMed]
5. Degrabriele, R. The Physiology of the Koala. *Sci. Am.* **1980**, *243*, 110–117.
6. Langman, V.A.; Rowe, M.; Forthman, D.; Whitton, B.; Langman, N.; Roberts, T.; Huston, K.; Boling, C.; Maloney, D. Thermal Assessment of Zoological Exhibits: Sea Lion enclosure at the Audubon Zoo. *Zoo Biol.* **1996**, *15*, 403–411. [CrossRef]
7. Warriss, P.D.; Pope, S.J.; Brown, S.N.; Wilkins, L.J.; Knowles, T.G. Estimating the body temperature of groups of pigs by thermal imaging. *Vet. Rec.* **2006**, *158*, 331–334. [CrossRef] [PubMed]
8. Geiser, F.; Westman, W.; Brownwyn, M.; McAllan, R.M.; Brigham, M.R. Development of thermoregulation and torpor in a marsupial: Energetic and evolutionary implications. *J. Comp. Physiol.* **2006**, *176*, 107–116. [CrossRef] [PubMed]
9. Johnson, S.R.; Rao, S.; Hussey, S.B.; Morley, P.S.; Traub-Dargatz, J.L. Thermographic Eye Temperature as an Index to Body Temperature in Ponies. *J. Equine Vet. Sci.* **2011**, *31*, 63–66. [CrossRef]
10. Zanghi, B.M. Eye and ear temperature using infared thermography are related to rectal temperature in dogs at rest or with exercise. *Front. Vet. Sci.* **2016**, *3*, 111. [CrossRef] [PubMed]
11. Barbieri, M.M.; Mclellan, W.A.; Wells, R.S.; Blum, J.E.; Hofmann, S.; Gannon, J.; Pabst, D.A. Using infrared thermography to assess seasonal trends in dorsal fin surface temperatures of free-swimming bottlenose dolphins (*Tursiops truncatus*) in Sarasota Bay, Florida. *Mar. Mamm. Sci.* **2010**, *26*, 53–66. [CrossRef]
12. Stevenson, R.D. The Relative Importance of Behavioural and Physiological Adjustments Controlling Body Temperature in Terrestrial Ecotherms. *Am. Nat.* **1985**, *126*, 362–386. [CrossRef]
13. Jackson, S. *Australian Mammals: Biology and Captive Management*; CSIRO Publishing: Melbourne, Australia, 2003.
14. Lee, A.; Martin, R. *The Koala: A Natural History*; New South Wales University Press: Sydney, Australia, 1988.
15. Foster, S.; Ijichi, C. The association between infrared thermal imagery of core eye temperature, personality, age and housing in cats. *Appl. Anim. Behav. Sci.* **2017**, *189*, 79–84. [CrossRef]
16. Wickins-Drazilova, D. Zoo Animal Welfare. *J. Agric. Environ. Ethics* **2005**, *19*, 27–36. [CrossRef]
17. Jerem, P.; Herborn, K.; McCafferty, D.; McKeegan, D.; Nager, R. Thermal Imaging to Study Stress Non-invasively in Unrestrained Birds. *J. Vis. Exp.* **2015**, *105*, 1–10. [CrossRef] [PubMed]
18. Stewart, M.; Webster, J.R.; Verkerk, G.A.; Schaefer, A.L.; Colyn, J.J.; Stafford, K.J. Non-invasive measurement of stress in dairy cows using infrared thermography. *Behav. Physiol.* **2007**, *92*, 520–525. [CrossRef] [PubMed]

Communication

Parasite Load and Site-Specific Parasite Pressure as Determinants of Immune Indices in Two Sympatric Rodent Species

Tim R. Hofmeester [1,†], Esther J. Bügel [1], Bob Hendrikx [1], Miriam Maas [2], Frits F. J. Franssen [2], Hein Sprong [2] and Kevin D. Matson [1,*]

1 Resource Ecology Group, Wageningen University, Droevendaalsesteeg 3a, 6708PB Wageningen, The Netherlands; tim.hofmeester@slu.se (T.R.H.); e.j.bugel@iclon.leidenuniv.nl (E.J.B.); b.hendrikx91@gmail.com (B.H.)

2 Centre for Infectious Disease Control, National Institute for Public Health and the Environment, Antonie van Leeuwenhoeklaan 9, 3721 MA Bilthoven, The Netherlands; miriam.maas@rivm.nl (M.M.); frits.franssen@rivm.nl (F.F.J.F.); hein.sprong@rivm.nl (H.S.)

* Correspondence: kevin.matson@wur.nl

† Present address: Department of Wildlife, Fish, and Environmental Studies, Swedish University of Agricultural Sciences, Skogsmarksgränd 7, 90183 Umeå, Sweden.

Received: 22 October 2019; Accepted: 20 November 2019; Published: 22 November 2019

Simple Summary: Wild animals can host diseases that affect humans (i.e., zoonotic diseases). However, not all wild animals are equal in their hosting capacities. In fact, the immune system, the main defense against diseases, varies within and among species. Within species, variation relates to two factors: parasite load and parasite pressure. Parasite load refers to the amount of parasites in or on an individual. Parasite pressure refers to the amount of parasites in a location. The importance of these factors in shaping the immune system of wild rodents, a group of animals known to host zoonotic diseases, is poorly understood. Overall, the rodent species we studied (bank voles and wood mice) hosted 5 microparasites, 9 ectoparasites, and 8 gastrointestinal parasites. We found that parasite load and parasite pressure related to different facets of the immune system. We also found that bank voles exhibited higher levels of one immune defense than wood mice, but higher levels of this defense correlated with a worm infection only in wood mice. Lastly, a white blood cell ratio correlated with infection by a zoonotic parasite. Studies like ours help to document the complexities of host–parasite interactions and how these interactions shape zoonotic disease risk in a changing world.

Abstract: Wildlife is exposed to parasites from the environment. This parasite pressure, which differs among areas, likely shapes the immunological strategies of animals. Individuals differ in the number of parasites they encounter and host, and this parasite load also influences the immune system. The relative impact of parasite pressure vs. parasite load on different host species, particularly those implicated as important reservoirs of zoonotic pathogens, is poorly understood. We captured bank voles (*Myodes glareolus*) and wood mice (*Apodemus sylvaticus*) at four sites in the Netherlands. We sampled sub-adult males to quantify their immune function, infestation load for ecto- and gastrointestinal parasites, and infection status for vector-borne microparasites. We then used regression trees to test if variation in immune indices could be explained by among-site differences (parasite pressure), among-individual differences in infestation intensity and infection status (parasite load), or other intrinsic factors. Regression trees revealed splits among sites for haptoglobin, hemagglutination, and body-mass corrected spleen size. We also found splits based on infection/infestation for haptoglobin, hemolysis, and neutrophil to lymphocyte ratio. Furthermore, we found a split between species for hemolysis and splits based on body mass for haptoglobin, hemagglutination, hematocrit, and body-mass corrected spleen size. Our results suggest that both parasite pressure and parasite load influence the immune system of wild rodents. Additional studies linking disease ecology and ecological immunology are needed to understand better the complexities of host–parasite interactions and how these interactions shape zoonotic disease risk.

Keywords: rodents; ecological immunology; natural antibodies; haptoglobin; neutrophil to lymphocyte ratio; immune strategy; vector-borne pathogens; parasitology; zoonosis

---

1. Introduction

Free-living wild animals are repeatedly exposed to different parasites. These parasites can be found on plants and other animals, in soil and water, and generally throughout the animal's environment [1,2]. When an animal interacts with one or more parasites, that animal's immune system can respond in different ways [3]. For example, some hosts engage in strategies of resistance, while others rely on tolerance [4]. These general strategies combined with myriad immunological defenses form a host's immunological phenotype [4,5]. Qualitative and quantitative differences in aspects of immunological phenotypes have been documented among individuals, populations, and species of animals through space and time [6–10]. As with other plastic phenotypic traits, immunological phenotypes and variation therein are shaped by evolutionary (i.e., genetic) and ecological (i.e., environmental) factors.

Two influential factors are parasite pressure and parasite load [1]. Parasite pressure is largely a characteristic of the broader environment, the specific habitat in which animals live, or both [1]. Exposure to parasite pressure across generations is thought to drive selection and shape immune system evolution, but this has mostly been tested indirectly [1,11]. Parasite load is a trait associated with individual animals [1,12]. The magnitude of parasite load likely depends on parasite pressure and other environmental characteristics, but critically, it also integrates host ecology and immunology [1,11]. Parasite load can be decomposed into several parameters, including the infection status (i.e., presence/absence) and infection intensity. In field studies of wild animals, parasite pressure, parasite load, and host immune defenses are rarely all characterized simultaneously in the same study population.

Knowledge is limited regarding the relative contribution of parasite pressure and parasite load to processes shaping immunological phenotypes in wild animals. Yet understanding these relationships is increasingly important in the light of emerging pathogens that can cause disease in humans [13]. These so-called zoonotic pathogens are often maintained in enzootic cycles by wildlife populations [14], and several zoonotic pathogens have increased in occurrence in recent decades [13]. Of the few studies that have begun to explore ecological immunology in the context of specific pathogens, even fewer study multiple zoonotic pathogens (e.g., [8]). Evaluating relationships among zoonotic pathogens and immune defenses helps in understanding these natural systems and the risks they pose to humans.

We quantified immunological and physiological indices and parasite load parameters in two common species of small rodent from wooded areas in the Netherlands (Supplementary Figure S1) that were previously shown to differ in their parasite pressure (Supplementary Table S1 [15]). Moreover, the bank vole (*Myodes glareolus*) and the wood mouse (*Apodemus sylvaticus*), our two study species, differ in their ability to *host* ticks and to *infect* ticks with zoonotic pathogens [6,16]. Bank voles, but not wood mice, acquire resistance to some ectoparasites (e.g., the tick *Ixodes ricinus* [16]), while wood mice mount a stronger antibody-mediated response against zoonotic pathogens than bank voles [6]. Given our interest in both parasite pressure and parasite load and given that infection with one parasite can mediate infection with another (i.e., via mechanisms of co-infection [17]), we took a holistic "parasite assemblage approach"; however, we also maintained a strong focus on vector-borne microparasites. To this end, we screened rodents for an array of ectoparasites, gastrointestinal parasites, and microparasites. (For a full list, see Supplementary Table S2.) We also characterized the immunological phenotypes of the same individuals via six indices of the immune system and other allied physiological systems.

We formulated two parasite-related hypotheses. First, if parasite pressure drives immunological phenotype, then populations from different sites are expected to express different immunological phenotypes. In general, higher parasite pressure is thought to select for stronger immune systems [1]. Second, if parasite load drives immunological phenotype, then individuals carrying higher parasite loads are expected to express immunological phenotypes that differ from those carrying lower loads, irrespective of population. The direction of this relationship likely depends on the parasite load parameter under consideration, since some members of the parasite assemblage can be immunostimulatory and others immunosuppressive [17]. Additionally, we expected intrinsic host factors to shape immunological phenotype. In the light of the differences between our study species described above, immunological indices are expected to correlate more strongly with microparasite infection status in wood mice compared to bank voles. Furthermore, immunological indices are expected to correlate positively with body mass (a proxy for age in rodents [18]), a result of immune system development.

2. Materials and Methods

2.1. Study Sites

Between 13 September and 7 October in 2016, we worked in four 1 ha wooded sites in the Netherlands: Buunderkamp, Herperduin, Maashorst, and Stameren (Supplementary Figure S1). Details about these sites, including exact locations, have been described previously [15]. We selected these specific sites based on a known gradient in tick burden on rodents (Supplementary Table S1 [15]) and based on spatial isolation to ensure independent populations of rodents and parasites (the closest neighboring sites, Herperduin and Maashorst, were separated by 5.5 km and a highway).

2.2. Rodent Trapping

In each study site, we established a 10 × 10 grid of trapping stations with 10 m between stations. With a pair of Longworth live traps (Heslinga Traps, Groningen, The Netherlands) per station, a grid consisted of 200 traps in total. We activated the traps at 20.00 h on four consecutive evenings and checked them the following mornings at 8.00 h. Traps were closed during the day. We baited the traps with mixed seeds, pieces of carrot and apple, and live mealworms; hay served as insulation. The mealworms were included as food for accidental bycatch of shrews (*Soricidae*). With their utilization of these resources, trapped animals showed no outward signs of stress upon removal from the traps, but any unmeasured effects of capture are expected to have been consistent across animals given the standardized trapping protocol.

Trapped rodents were transferred into a transparent plastic bag to allow for species identification with minimal handling. Once identified, any nontarget species were immediately released. We visually sexed all bank voles and wood mice and then released any females after marking them by fur clipping [19] to facilitate release if recaptured. We weighed all males and evaluated testes development; individuals >20 g or with visible testes were marked and released. By targeting sub-adult males that were not sexually active, we aimed to reduce variation in immunological parameters due to age or reproductive state [2].

2.3. Rodent Anesthetization, Sample Collection and Handling

All target males were anaesthetized with a 0.15–0.25 mL injection (depending on body mass) of a 1:2 mixture of xylazine (2 mg/mL) and ketamine (10 mg/mL). After anesthetization, blood was collected via eye extraction, and the animals were euthanized by cervical dislocation. Rodent trapping, anesthetization, euthanization, and all other aspects of the animal experiments were approved in 2016 or earlier by the Central Committee Animal Experimentation in the Netherlands (AVD104002016546), the Animal Welfare Body of Wageningen University (WUR-2016044), and The Netherlands Ministry of Economic Affairs (FF/75A/2013/003).

Blood was collected directly into a tube (MiniCollect® Tube 0.8 mL Z Serum Sep Clot Activator, 450472; Greiner Bio-One B.V., Alphen aan den Rijn, The Netherlands), and samples were kept in a cool box with ice in the field until processing in the lab later the same day. Samples were centrifuged and serum was collected in accordance with the manufacturer's instructions. After centrifugation, serum was transferred to a fresh microcentrifuge tube and frozen at −20 °C until laboratory analyses, which took place from late October to mid-November 2016.

At the moment of blood collection, we made a blood smear per individual using a drop of the fresh whole blood. Smears were fixed in absolute methanol in the field and stained (Modified Giemsa Stain, GS500-500 ML; Sigma-Aldrich Inc., St. Louis, MO, USA) later in the lab. We also filled one heparinized capillary tube per individual, which was centrifuged at 12,000 rpm for 10 min. From this, we recorded hematocrit values per individual.

From the carcasses, we collected kidneys, liver, spleen, and lungs and some ear tissue in the field for purposes of microparasite screening. The lungs were placed in RNAlater (Thermo Fisher Scientific, Waltham, MA, USA). We kept these samples in a cool box with ice until later the same day, when the lungs were stored at 4 °C and all other organs and tissues and the remaining carcass were stored at −20 °C. Spleen size (mass in mg) was recorded on the day of collection before being frozen. After 3–5 days, we removed the lung samples from the RNAlater and stored them individually in microcentrifuge tubes at −20 °C with the rest of the samples.

2.4. Ectoparasite Screening

We screened all carcasses for arthropod ectoparasites. Ectoparasites were collected by combing the fur, checking the skin, and carefully inspecting each carcass's storage bag. We also collected and stored any ectoparasites that left the carcass during the dissection process. Ectoparasites were stored in 70% ethanol in microcentrifuge tubes (one tube per parasite species per host individual) at −20 °C.

Ectoparasites were counted per species and life stage. Tick and flea species were identified using established identification keys [20,21]. Lice and mite species were identified by Herman J.W.M. Cremers (Utrecht University, Utrecht, The Netherlands), who also confirmed flea species identity.

2.5. Gastrointestinal Parasite Screening

Gastrointestinal parasites were isolated following previously described protocols [22]. Briefly, mouse intestines were separated individually into small (duodenum up to ileum) and large intestine (including caecum and colon), and their contents were homogenized in phosphate buffered saline. Parasites were collected by sieving the suspension over a 63 µm mesh size sieve. The stomach was screened for parasites macroscopically and microscopically. Isolated parasites were counted, sexed, identified morphologically, and stored in 70% ethanol.

2.6. Screening for Microparasites

DNA and RNA were extracted from tissue samples in a diagnostic laboratory using a robot (MagNA Pure Compact Extraction Robot; Roche, Basel, Switzerland) and Nucleic Acid Isolation Kit I (Roche, Roche, Basel, Switzerland) following the manufacturer's instructions. To detect potential cross-contamination, we included negative controls in each extraction batch. To minimize contamination and false positives, all main steps (extraction, PCR mix preparation, sample addition, and (q)PCR analyses) were performed in separate air-locked dedicated labs.

We analyzed all nucleic acid extractions with different (multiplex) qPCRs carried out on a LightCycler 480 (Roche Diagnostics Nederland B.V, Almere, The Netherlands). These qPCRs were performed as described previously and based on gene fragments specific for the (vector-borne) microparasites of interest: *Borrelia burgdorferi* s. l. (two targets, [23]), *Borrelia miyamotoi* [24], *Anaplasma phagocytophilum* [25,26], *Candidatus* Neoehrlichia mikurensis [27], *Rickettsia* spp. [28], *Leptospira* spp. [29], and *Bartonella* spp. [30]. We also analyzed the nucleic acid extractions for the presence of four viruses: tick-borne encephalitis virus, Tula hantavirus, Puumala hantavirus, and Eyach virus. For the first three, we performed reverse transcription real-time PCRs as previously described [31–34]. Positive (plasmid) controls and negative (water) controls were used on every plate.

We used methodology that has not been previously described for two microparasites: *Spiroplasma ixodetes* and Eyach virus. We screened for *S. ixodetes* using exactly the same conditions as for *B. burgdorferi* s. l. but with (rpoBF) 5′-TGTTGGACCAAACGAAGTTG-3′ and (rpoBR) 5′-CCAACAATTGGTGTTTGGGG-3′ as primers and (rpoBP) 5′-(FAM)GCTAACCGTGCTTT AATGGG(BHQ1)-3′ as the probe. For the Eyach virus, we performed a reverse transcription real-time PCR targeting the VP2 of the Eyach virus genome with (EyachF) 5′-TGGCTGACAAC ATGACGGATA-3′ and (EyachR) 5′-GGCCTCACGATACTTTCGATT-3′ as primers and (EyachP) 5′-ACGGGCTCGGTACTTCGGTTGAGAT-3′ as the probe. We used 20 μL with TaqMan Fast Virus 1-Step Master Mix (Thermo Fisher Scientific, Waltham, MA, USA), 5 μL of sample, and 0.2 μM for the primers and probes for the qPCR, which was performed with a 20 min reverse transcription step at 50 °C, denaturation at 95 °C for 30 s, and 50 cycles of 95 °C for 10 s and 60 °C for 30 s.

Samples positive for *Bartonella* spp. were subjected to conventional PCR and sequencing to determine species identity [35,36]. *Bartonella* sequences were compared with reference sequences from Genbank using the unweighted pair group method with arithmetic mean-based (UPGMA) hierarchical clustering.

2.7. Immunological Assays

2.7.1. Haptoglobin

Using a commercially available assay kit (Tridelta PHASE Haptoglobin Assay, TP-801; Tridelta Development Limited, Maynooth, County Kildare, Ireland), we measured circulating concentrations of haptoglobin, which is an acute phase protein that increases in concentration in response to infection, inflammation, or trauma [7,10]. We followed the manufacturer's instructions, measuring absorbance at 630 nm 5 min after initiation of the color change reaction. We calculated concentrations in our samples using a linear standard curve that we generated from six standards ranging from 0.039 to 1.25 mg Ml^{-1}. Most serum samples were analyzed in singlicate, but samples from four individuals were run in duplicate to allow for the quantification of within assay (i.e., within plate) variability. The average coefficient of variation of the four duplicated samples was 5.6%. For each duplicated sample, we used the mean concentration in further analyses.

2.7.2. Hemolysis and Hemagglutination

We measured titers of both complement-driven hemolysis and natural antibody-driven hemagglutination following standard protocols [7,9]. Since the test serum was from mammals, we used blood from a bird (*Columba livia domestica*) to make the required 1% suspension of exogenous erythrocytes. All test samples were analyzed in singlicate, but each of the six assay plates included a duplicate (and in one case, triplicate) sample from a brown rat (*Rattus norvegicus*) to allow for the quantification of within assay (i.e., within plate) and among assay (i.e., among plate) variability. On average, the rat standard exhibited lysis of 4.4 titers and agglutination of 6.1 titers, and the variability (coefficient of variation) was as follows: hemolysis, within 12.4%; hemolysis, among 12.5%; hemagglutination, within 8.3%; and hemagglutination, among 15.5%.

2.7.3. Neutrophil to Lymphocyte Ratio

One author (E.B.) analyzed all of the stained blood smears using a light microscope (at 1000×). Per slide, she counted the neutrophils and lymphocytes until a combined total of up to 100 cells was achieved ($\overline{x}_{total}$ = 83 cells). This process was repeated twice per slide, and the average count of each cell type was used in further analyses of their ratio (neutrophil to lymphocyte ratio).

2.8. Data Analysis

We explored relationships between our six response variables related to the immune system and allied physiological systems (haptoglobin concentration (mg mL^{-1}), hemolysis (titers), hemagglutination (titers), hematocrit (%), body-mass corrected spleen size (mg g^{-1}), and neutrophil to lymphocyte ratio) and several explanatory variables using regression trees [37,38]. We included the following explanatory variables per host individual: site, ecto- and gastrointestinal parasite infestation intensities (i.e., numbers of individuals of each parasite species), microparasite infection statuses (i.e., infected or not with each parasite species, based on molecular techniques), host species identity, and body mass.

Regression trees allowed us to explore the most parsimonious splits in the data using a nonparametric approach that does not assume linearity or independence of data. Furthermore, this analytical approach is ideal for exploring of patterns without a priori hypotheses for specific parasite species. We fitted regression trees in R 3.5.1 [39] using the *rpart* package (v 4.1.13 [40]). Regression trees can be read as a decision tree, where the data are split into two groups at each node. The regression tree for each immunological or physiological response variable was built by recursively partitioning data using an algorithm to split the data into two groups based on the best predictor. This process is repeated for each of the newly formed groups separately, maximizing the deviance in the response variable. A cross-validation step was employed to prune trees using the complexity parameter that accounts for tree complexity and the variance explained. This step resulted in the smallest possible trees with minimum classification error. In contrast to many conventional statistical methods, most regression tree analyses do not calculate statistical significance or *p*-values. To accommodate readers in their interpretation, while adhering to the nonparametric nature of the regression trees, we performed independent 2-group Mann–Whitney U tests [41] to test for differences between groups at each node.

3. Results

We sampled 36 rodents (10 bank voles and 26 wood mice) from the four locations (full dataset available in Supplementary Table S2). We identified five vector-borne microparasites using qPCR and sequencing: *Bartonella grahamii, Bartonella taylorii, Borrelia burgdorferi* s.l., *Borrelia miyamotoi,* and *Candidatus* Neoehrlichia mikurensis (Table 1). We found nine ectoparasite species: two tick species (*Ixodes ricinus* and *Ixodes trianguliceps*), four mite species (*Echinonyssus isabellinus, Eulaelaps stabularis, Haemogamasus nidi,* and *Laelaps agilis*), two flea species (*Ctenophtalmus agyrtes* and *Megabothris turbidus*), and one louse species (*Polyplax serrata*; Table 1). We also found eight gastrointestinal-parasite species: *Aonchoteca murissylvatici, Aspiculuris tianjinensis, Heligmosomoides polygyrus, Heterakis spumosa, Hymenolepis diminuta, Syphacia petruzewiczi, Syphacia stroma,* and an unidentified species of coiled nematode that was not *Trichinella* (Table 1). All serological tests were negative.

Table 1. Overview of infection with vector-borne microparasites and infestation with ecto- and gastrointestinal parasites in wood mice and bank voles in four Dutch forest sites.

Parasite	Bank Vole (n = 10)		Wood Mouse (n = 26)	
Microparasites				
Bartonella grahamii [1]	0.40		0.04	
Bartonella taylorii [1]	0.10		0.31	
Borrelia burgdorferi s.l. [1]	0		0.04	
Borrelia miyamotoi [1]	0.10		0.23	
Candidatus Neoehrlichia mikurensis [1]	0.30		0.12	
Ectoparasites				
Ixodes ricinus (larvae) [2]	3.20 (0–8)	[0.80]	11.08 (0–80)	[0.92]
Ixodes ricinus (nymphs) [2]	0	[0]	0.08 (0–2)	[0.04]
Ixodes trianguliceps (larvae) [2]	0.70 (0–4)	[0.40]	0.38 (0–4)	[0.15]
Echinonyssus isabellinus [2]	0.10 (0–1)	[0.10]	0	[0]
Eulaelaps stabularis [2]	0	[0]	1.23 (0–27)	[0.19]
Haemogamasus nidi [2]	0	[0]	0.35 (0–9)	[0.04]
Laelaps agilis[2]	0	[0]	3.69 (0–27)	[0.58]
Ctenophtalmus agyrtes [2]	0.10 (0–1)	[0.10]	0.08 (0–1)	[0.08]
Megabothris turbidus [2]	0	[0]	0.08 (0–2)	[0.04]
Polyplax serrata [2]	0	[0]	0.85 (0–8)	[0.19]
Gastrointestinal parasites				
Aonchoteca murissylvatici [2]	2.10 (0–14)	[0.30]	5.65 (0–75)	[0.35]
Aspiculuris tianjinensis [2]	15.40 (0–150)	[0.20]	2.31 (0–29)	[0.19]
Heligmosomoides polygyrus [2]	2.80 (0–18)	[0.40]	0.38 (0–3)	[0.31]
Heterakis spumosa [2]	0	[0]	0.04 (0–1)	[0.04]
Hymenolepis diminuta [2]	0.10 (0–1)	[0.10]	0	[0]
Syphacia petruzewiczi [2]	3.80 (0–24)	[0.20]	0	[0]
Syphacia stroma [2]	0	[0]	12.46 (0–76)	[0.58]
Non-*Trichinella* coiled nematode [2]	0.90 (0–7)	[0.20]	0	[0]

[1] Prevalence of (vector-borne) microparasites given as ratio of total number of animals with infection. [2] Average parasite load as well as prevalence (between square brackets) given as mean and range (in parentheses) of burden and ratio of total number of animals with infestation, respectively.

We found that simple trees with 3–4 nodes best described the data for all immunological and physiological indices. These trees revealed both differences among sites and differences in relation to parasite infestation or infection. Haptoglobin concentration was highest in animals with an *Ixodes trianguliceps* infestation (Mann–Witney U test, $p = 0.035$) and second highest in animals from Herperduin without *I. trianguliceps* ($p = 0.022$, Figure 1A). Haptoglobin concentration was lowest in animals with a body mass ≥16.5 gram in the other sites ($p = 0.045$, Figure 1A). Hemolysis titer was highest in bank voles ($p = 0.00093$) and intermediate in wood mice with a *Heligmosomoides polygyrus* infection ($p = 0.050$, Figure 1B). Hemagglutination titer was lowest in rodents <15.5 gram ($p = 0.011$) and highest in rodents from three sites (Buunderkamp, Herperduin and Stameren; $p = 0.033$, Figure 1C). Neutrophil to lymphocyte ratio was highest in animals infected with *Borrelia miyamotoi* ($p = 0.0066$) and lowest in animals infested with ≥3 *Laelaps agilis* ($p = 0.12$, Figure 1D). Hematocrit was lowest in rodents <13.5 gram ($p = 0.0099$) and highest in rodents between 13.5 and 17.5 gram ($p = 0.11$, Figure 1E). Body-mass corrected spleen size was highest in animals ≥17.5 gram ($p = 0.0057$) and lowest in animals from Buunderkamp, Herperduin and Maashorst ($p = 0.036$, Figure 1F).

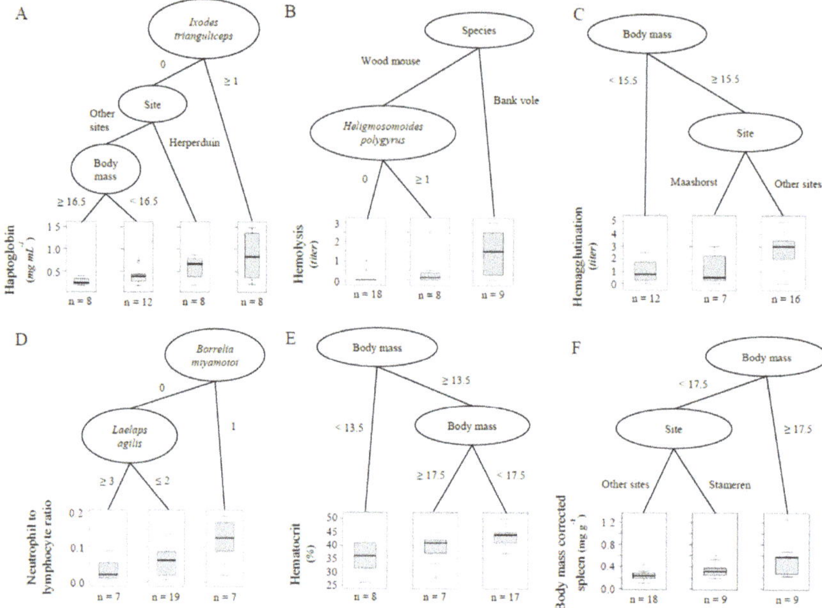

Figure 1. Regression tree showing the most parsimonious split for (**A**) haptoglobin, (**B**) hemolysis, (**C**) hemagglutination, (**D**) neutrophil to lymphocyte ratio, (**E**) hematocrit, and (**F**) body-mass corrected spleen size. Boxplots show the spread of values (median and quantiles) within each group identified by the different splits in the tree. Sample size for each group is given underneath the boxplots. Body mass is given in grams.

4. Discussion

Evaluating relationships between zoonotic pathogens and reservoir hosts, particularly in the context of host immune defenses, can provide new insights into these natural systems and the risks posed to humans. The immunological phenotype of hosts is shaped by both parasite pressure and parasite load [1]. To help to disentangle these two factors, we quantified immunological and related physiological indices and parasite load parameters in bank voles and wood mice from wooded sites in the Netherlands varying in parasite pressure. We found immunological and physiological differences between groups that were and were not infected or infested with parasites, but we also found differences among sites. Defining the biological meaning of such differences, even when using some of the most widely applied assays of ecological immunology, remains challenging, but our results offer some new context to the measured indices. For example, interspecific differences in hemolysis were larger than intraspecific differences related to infestation (parasite load), supporting an earlier idea that this index is relatively invariable in relation to current health status, despite the taxonomic variation it displays [7,9]. In addition, both infestation (parasite load) and site (parasite pressure) associated differences were related to haptoglobin values outside of clinically normal ranges (rodent: 0.25–0.51 mg mL^{-1}, murine: 0.00–0.10 mg mL^{-1}; both reported by the assay manufacturer). Thus, individuals of both rodent species that were either feeding *I. trianguliceps* ticks or living at Herperduin can be viewed as enduring a type of systemic innate immune response, known as an acute phase response [42]). Overall, we can safely conclude that the measured immunological defenses are shaped by a combination of current parasite load and differences in parasite pressure among spatially distinct sites.

We hypothesized that if parasite pressure drives immunological phenotype, then populations from sites that differ in parasite pressure should express different levels of the measured immune indices. In general, higher parasite pressure is thought to select for stronger immune systems [1].

We found partial support for our hypothesis in the form of differences among sites in terms of haptoglobin, hemagglutination, and body-mass corrected spleen size. In all three cases, one site differed from the other three. This suggests that circumstances specific to some sites might influence parasite pressure (e.g., which parasite species and pathogen strains successfully colonized a site, the overall parasite and pathogen richness or diversity), evolutionary responses to that pressure (e.g., host population genetics), or both. However, the patterns reported here are inconsistent with previously reported differences in parasite pressure, which was characterized by densities of *Ixodes ricinus* and prevalences of several tick-borne pathogens [15]. For example, sites with higher levels of haptoglobin (Herperduin), hemagglutination (Buunderkamp, Stameren and Herperduin), and body-mass corrected spleen size (Stameren) were not the same as the site with the highest parasite pressure in 2013–2014 (Buunderkamp). We did find that endoparasite loads were highest in rodents from Herperduin, which could explain the higher level of haptoglobin at this site, suggesting that a more thorough approach is needed to quantify parasite pressure at a site [2]. Even though sites were on average 30 km apart and were previously shown to have clear differences in tick densities and tick-borne microparasite prevalences [15], these distinctions might not have consistently translated to different immunological phenotypes in measurable ways. Previous studies showing between-site differences in host immunology employed study designs with extreme environmental differences [43].

We also found support for our second hypothesis: individuals burdened with higher parasite loads were expected to differ from those carrying lower loads in terms of the measured immune indices. The direction of effect likely depends on the parasite type (immunostimulatory vs. immunosuppressive) and immune index under consideration. Indeed, we found that the highest and lowest levels of haptoglobin and the neutrophil to lymphocyte ratio were associated with infection or infestation. Concentrations of haptoglobin were higher in individuals infested with the tick *Ixodes trianguliceps*. Haptoglobin is a positive acute phase protein [44], meaning concentrations increase in response to infection, inflammation, or trauma [42], any of which could result from the mouthparts of a tick puncturing the skin of a host and the multiday feeding period that follows [45]. We also found the highest neutrophil to lymphocyte ratios in animals infected with *B. miyamotoi*. This result could be caused by increased neutrophil production or decreased lymphocyte numbers in these infected animals, but neither of these effects were seen in a case-study of a human patient [46]. Furthermore, we found the lowest neutrophil to lymphocyte ratios in animals infested with *Laelaps agilis*. This species of mite is known to be a vector of *Hepatozoon* spp., blood parasites that can infect wild rodents, but for which we did not test our samples [47]. Since *Hepatozoon* spp. can infect leukocytes in rodents [48,49], such an infection might be a mechanism behind the mite infestation effect; however, in other animals (e.g., frogs [50]), neutrophil to lymphocyte ratio was not correlated with infection intensity of *Hepatozoon* spp. Nevertheless, since increased neutrophil to lymphocyte ratios are often associated with stressed or diseased states [51], the lower values we observed in mite infested individuals suggest another process at work (e.g., possible immunosuppression).

Finally, we found differences related to species identity and body mass. Hemolysis was the only immune index for which we found a difference between the two study species: Overall bank voles exhibited higher titers than wood mice. Furthermore and tangentially related to our a priori expectation, hemolysis was the only immune index for which we found an infection-related difference between the two study species. Hemolysis titers were higher in *Heligmosomoides polygyrus* (helminth) infected wood mice but not in similarly infected bank voles. *Heligmosomoides polygyrus*, a commonly used model of helminth infections, is known to regulate immune function in laboratory mice [52]. To our knowledge, our study is the first to show immunological differences associated to natural infection with *H. polygyrus*. It must be noted, however, that the effects of microparasite infection status never showed this type of species dependence. We also found that four immune indices showed three different types of relationships with body mass. Hemagglutination and body-mass corrected spleen size related positively with body mass; haptoglobin related negatively with body mass; and hematocrit showed an optimum at middle body masses. While the effects of age on immune function in rodents are known, if not fully understood [2], our results highlight the complexity of these dynamics. The immune system in wild rodents does not simply "mature" as individuals grow heavier (and older), even if some individual components, such as hemagglutination, show such a pattern here and elsewhere [7,9]. Instead, the observed relationships hint at an influence of body condition or composition, but this potential mechanism could not be investigated given our lack of data on the structural size of individuals [18].

Overall, our findings offer new insights into relationships between specific parasites and immunological and physiological indices, as well as broader differences between species and among sites. Notably, differences related to study site and infection status seemed to exert a greater impact on immune phenotype than host species identity, even though the two rodent study species are thought to play different roles in the maintenance and transmission of tick-borne zoonotic pathogens [6,16]. While our results document influential roles for both the environmental characteristic of parasite pressure and the organismal characteristic of parasite load on the immunological phenotype of wild animals, additional studies linking disease ecology and ecological immunology are needed to better understand the complexities of how host–parasite interactions play out through space and time in different environments. For example, repeated sampling of sites over longer (i.e., multiyear) time periods would be invaluable for characterizing the potential for parasite pressure to drive immune system evolution. Lastly, our study illustrates the possibility and added value of using a holistic approach targeting diverse parasites and multiple aspects of host immunology and physiology when investigating zoonotic pathogens and their vectors and reservoirs [53]. The resulting insights, such as overall differences between species (e.g., hemolysis) and interspecific similarities and differences in the immunomodulatory effects of infection (e.g., neutrophil to lymphocyte ratio and hemolysis, respectively), can help to shape new questions, for example, about host competence [53]. In this study, our focus was on understanding immunological variation, but a similar analytical approach could be used to explore variation in the burden (e.g., infection intensity) of one or more parasites.

Supplementary Materials: The following are available online at http://www.mdpi.com/2076-2615/9/12/1015/s1, Supplementary Table S1: Site characteristics, and Supplementary Table S2: Full dataset. Supplementary Figure S1: Map with locations of study sites. Left: map of the Netherlands with location of detailed map. Right: detailed map with location of the four study sites.

Author Contributions: Conceptualization, T.R.H., E.J.B., B.H., H.S., and K.D.M.; methodology, T.R.H., E.J.B., B.H., M.M., F.F.J.F., H.S., and K.D.M.; data analysis, T.R.H., E.J.B., and B.H.; data curation, T.R.H.; writing—original draft preparation, E.J.B. and B.H.; writing—final draft preparation, T.R.H. and K.D.M.; writing—review and editing, T.R.H., E.J.B., B.H., M.M., F.F.J.F., H.S., and K.D.M.; data visualization, T.R.H.; supervision, T.R.H. and K.D.M.; project administration, T.R.H. and K.D.M.; funding acquisition, T.R.H., M.M., F.F.J.F, H.S., and K.D.M.

Funding: This research was funded by the de Vos Vector Borne Diseases Fund at Wageningen University and by the Dutch Ministry of Health, Welfare, and Sport (VWS).

Acknowledgments: We are grateful to Herman J.W.M. Cremers (Utrecht University) for his help with ectoparasite species identification and to Marieke de Swart (National Institute for Public Health and Environment) for her help with microparasite screening. We also thank Ger de Vries Reilingh, Joop A.J. Arts, Rudie E. Koopmanschap, Henk K. Parmentier and the rest of the Adaptation Physiology Group (Wageningen University) for granting us access to their laboratory and for technical assistance.

Conflicts of Interest: The authors declare no conflict of interest. The funders had no role in the design of the study; in the collection, analyses, or interpretation of data; in the writing of the manuscript; or in the decision to publish the results.

References

1. Horrocks, N.P.C.; Matson, K.D.; Tieleman, B.I. Pathogen Pressure Puts Immune Defense into Perspective. *Integr. Comp. Biol.* **2011**, *51*, 563–576. [CrossRef] [PubMed]
2. Viney, M.; Riley, E.M. The Immunology of Wild Rodents: Current Status and Future Prospects. *Front. Immunol.* **2017**, *8*, 1481. [CrossRef] [PubMed]
3. Råberg, L.; Graham, A.L.; Read, A.F. Decomposing health: Tolerance and resistance to parasites in animals. *Philos. Trans. R. Soc. B Biol. Sci.* **2009**, *364*, 37–49. [CrossRef] [PubMed]

4. Baucom, R.S.; de Roode, J.C. Ecological immunology and tolerance in plants and animals. *Funct. Ecol.* **2011**, *25*, 18–28. [CrossRef]
5. Martin, L.B.; Hawley, D.M.; Ardia, D.R. An introduction to ecological immunology. *Funct. Ecol.* **2011**, *25*, 1–4. [CrossRef]
6. Kurtenbach, K.; Dizij, A.; Seitz, H.M.; Margos, G.; Moter, S.E.; Kramer, M.D.; Wallich, R.; Schaible, U.E.; Simon, M.M. Differential immune responses to Borrelia burgdorferi in European wild rodent species influence spirochete transmission to Ixodes ricinus L. (Acari: Ixodidae). *Infect. Immun.* **1994**, *62*, 5344–5352.
7. Matson, K.D.; Cohen, A.A.; Klasing, K.C.; Ricklefs, R.E.; Scheuerlein, A. No simple answers for ecological immunology: Relationships among immune indices at the individual level break down at the species level in waterfowl. *Proc. R. Soc. B-Biol. Sci.* **2006**, *273*, 815–822. [CrossRef]
8. Previtali, M.A.; Ostfeld, R.S.; Keesing, F.; Jolles, A.E.; Hanselmann, R.; Martin, L.B. Relationship between pace of life and immune responses in wild rodents. *Oikos* **2012**, *121*, 1483–1492. [CrossRef]
9. Matson, K.D.; Ricklefs, R.E.; Klasing, K.C. A hemolysis-hemagglutination assay for characterizing constitutive innate humoral immunity in wild and domestic birds. *Dev. Comp. Immunol.* **2005**, *29*, 275–286. [CrossRef]
10. Matson, K.D.; Horrocks, N.P.C.; Versteegh, M.A.; Tieleman, B.I. Baseline haptoglobin concentrations are repeatable and predictive of certain aspects of a subsequent experimentally-induced inflammatory response. *Comp. Biochem. Physiol. A Mol. Integr. Physiol.* **2012**, *162*, 7–15. [CrossRef]
11. Moyer, B.R.; Drown, D.M.; Clayton, D.H. Low humidity reduces ectoparasite pressure: Implications for host life history evolution. *Oikos* **2002**, *97*, 223–228. [CrossRef]
12. Rynkiewicz, E.C.; Hawlena, H.; Durden, L.A.; Hastriter, M.W.; Demas, G.E.; Clay, K. Associations between innate immune function and ectoparasites in wild rodent hosts. *Parasitol. Res.* **2013**, *112*, 1763–1770. [CrossRef]
13. Jones, K.E.; Patel, N.G.; Levy, M.A.; Storeygard, A.; Balk, D.; Gittleman, J.L.; Daszak, P. Global trends in emerging infectious diseases. *Nature* **2008**, *451*, 990–993. [CrossRef] [PubMed]
14. Daszak, P.; Cunningham, A.A.; Hyatt, A.D. Emerging Infectious Diseases of Wildlife - Threats to Biodiversity and Human Health. *Science* **2000**, *287*, 443–449. [CrossRef] [PubMed]
15. Hofmeester, T.R.; Jansen, P.A.; Wijnen, H.J.; Coipan, E.C.; Fonville, M.; Prins, H.H.T.; Sprong, H.; van Wieren, S.E. Cascading effects of predator activity on tick-borne disease risk. *Proc. R. Soc. B* **2017**, *284*. [CrossRef] [PubMed]
16. Dizij, A.; Kurtenbach, K. *Clethrionomys glareolus*, but not *Apodemus flavicollis*, acquires resistance to *Ixodes ricinus* L., the main European vector of *Borrelia burgdorferi*. *Parasite Immunol.* **1995**, *17*, 177–183. [CrossRef]
17. Telfer, S.; Lambin, X.; Birtles, R.; Beldomenico, P.; Burthe, S.; Paterson, S.; Begon, M. Species interactions in a parasite community drive infection risk in a wildlife population. *Science* **2010**, *330*, 243–246. [CrossRef]
18. Campbell, M.T.; Dobson, F.S. Growth and Size in Meadow Voles (*Microtus pennsylvanicus*). *Am. Midl. Nat.* **1992**, *128*, 180–190. [CrossRef]
19. Powell, R.A.; Proulx, G. Trapping and marking terrestrial mammals for research: Integrating ethics, performance criteria, techniques, and common sense. *ILAR J.* **2003**, *44*, 259–276. [CrossRef]
20. Hillyard, P.D. *Ticks of North-West Europe*; Backhuys Publishers: London, UK, 1996.
21. Whitaker, A.P. *Handbooks for the Identification of British Insects: Fleas (Siphonaptera)*, 2nd ed.; Royal Entomological Society: London, UK, 2007.
22. Franssen, F.; Swart, A.; van Knapen, F.; van der Giessen, J. Helminth parasites in black rats (*Rattus rattus*) and brown rats (*Rattus norvegicus*) from different environments in the Netherlands. *Infect. Ecol. Epidemiol.* **2016**, *6*, 31413. [CrossRef]
23. Heylen, D.; Matthysen, E.; Fonville, M.; Sprong, H. Songbirds as general transmitters but selective amplifiers of Borrelia burgdorferi sensu lato genotypes in Ixodes rinicus ticks. *Environ. Microbiol.* **2014**, *16*, 2859–2868. [CrossRef] [PubMed]
24. Hovius, J.W.R.; De Wever, B.; Sohne, M.; Brouwer, M.C.; Coumou, J.; Wagemakers, A.; Oei, A.; Knol, H.; Narasimhan, S.; Hodiamont, C.J.; et al. A case of meningoencephalitis by the relapsing fever spirochaete Borrelia miyamotoi in Europe. *Lancet* **2013**, *382*, 658. [CrossRef]
25. Courtney, J.W.; Kostelnik, L.M.; Zeidner, N.S.; Massung, R.F. Multiplex real-time PCR for detection of *Anaplasma phagocytophilum* and *Borrelia burgdorferi*. *J. Clin. Microbiol.* **2004**, *42*, 3164–3168. [CrossRef] [PubMed]

26. Jahfari, S.; Coipan, E.C.; Fonville, M.; van Leeuwen, A.; Hengeveld, P.; Heylen, D.; Heyman, P.; van Maanen, C.; Butler, C.M.; Földvári, G.; et al. Circulation of four Anaplasma phagocytophilum ecotypes in Europe. *Parasit. Vectors* **2014**, *7*, 365. [CrossRef]
27. Jahfari, S.; Fonville, M.; Hengeveld, P.; Reusken, C.; Scholte, E.J.; Takken, W.; Heyman, P.; Medlock, J.M.; Heylen, D.; Kleve, J.; et al. Prevalence of Neoehrlichia mikurensis in ticks and rodents from North-west Europe. *Parasit. Vectors* **2012**, *5*, 74. [CrossRef]
28. Stenos, J.; Graves, S.R.; Unsworth, N.B. A highly sensitive and specific real-time PCR assay for the detection of spotted fever and typhus group Rickettsiae. *Am. J. Trop. Med. Hyg.* **2005**, *73*, 1083–1085. [CrossRef]
29. Ahmed, A.; Engelberts, M.F.M.; Boer, K.R.; Ahmed, N.; Hartskeerl, R.A. Development and validation of a real-time PCR for detection of pathogenic leptospira species in clinical materials. *PLoS ONE* **2009**, *4*, e7093. [CrossRef]
30. Müller, A.; Reiter, M.; Schötta, A.M.; Stockinger, H.; Stanek, G. Detection of *Bartonella* spp. in *Ixodes ricinus* ticks and *Bartonella* seroprevalence in human populations. *Ticks Tick-Borne Dis.* **2016**, *7*, 763–767.
31. Lindblom, P.; Wilhelmsson, P.; Fryland, L.; Sjöwall, J.; Haglund, M.; Matussek, A.; Ernerudh, J.; Vene, S.; Nyman, D.; Andreassen, A.; et al. Tick-borne encephalitis virus in ticks detached from humans and follow-up of serological and clinical response. *Ticks Tick-Borne Dis.* **2014**, *5*, 21–28. [CrossRef]
32. Kramski, M.; Meisel, H.; Klempa, B.; Krüger, D.H.; Pauli, G.; Nitsche, A. Detection and typing of human pathogenic hantaviruses by real-time reverse transcription-PCR and pyrosequencing. *Clin. Chem.* **2007**, *53*, 1899–1905. [CrossRef]
33. Maas, M.; de Vries, A.; van Roon, A.; Takumi, K.; van der Giessen, J.; Rockx, B. High Prevalence of Tula Hantavirus in Common Voles in The Netherlands. *Vector Borne Zoonotic Dis. Larchmt. N* **2017**, *17*, 200–205. [CrossRef] [PubMed]
34. De Vries, A.; Vennema, H.; Bekker, D.L.; Maas, M.; Adema, J.; Opsteegh, M.; van der Giessen, J.W.B.; Reusken, C.B.E.M. Characterization of Puumala hantavirus in bank voles from two regions in the Netherlands where human cases occurred. *J. Gen. Virol.* **2016**, *97*, 1500–1510. [CrossRef] [PubMed]
35. Norman, A.F.; Regnery, R.; Jameson, P.; Greene, C.; Krause, D.C. Differentiation of Bartonella-like isolates at the species level by PCR-restriction fragment length polymorphism in the citrate synthase gene. *J. Clin. Microbiol.* **1995**, *33*, 1797–1803. [PubMed]
36. Tijsse-Klasen, E.; Fonville, M.; Gassner, F.; Nijhof, A.M.; Hovius, E.K.E.; Jongejan, F.; Takken, W.; Reimerink, J.R.; Overgaauw, P.A.M.; Sprong, H. Absence of zoonotic Bartonella species in questing ticks: First detection of Bartonella clarridgeiae and Rickettsia felis in cat fleas in the Netherlands. *Parasit. Vectors* **2011**, *4*, 61. [CrossRef]
37. De'ath, G.; Fabricius, K.E. Classification and regression trees: A powerful yet simple technique for ecological data analysis. *Ecology* **2000**, *81*, 3178–3192. [CrossRef]
38. Krzywinski, M.; Altman, N. Points of Significance: Classification and regression trees. *Nat. Methods* **2017**, *14*, 757–758. [CrossRef]
39. R Core Team. *R: A Language and Environment for Statistical Computing*; R Foundation for Statistical Computing: Vienna, Austria, 2019.
40. Therneau, T.; Atkinson, B.; Ripley, B. *Rpart: Recursive Partitioning and Regression Trees*; Mayo Foundation: Rochester, MN, USA, 2019; Available online: https://cran.r-project.org/web/packages/rpart/index.html (accessed on 2 October 2019).
41. Mann, H.B.; Whitney, D.R. On a Test of Whether one of Two Random Variables is Stochastically Larger than the Other. *Ann. Math. Stat.* **1947**, *18*, 50–60. [CrossRef]
42. Quaye, I.K. Haptoglobin, inflammation and disease. *Trans. R. Soc. Trop. Med. Hyg.* **2008**, *102*, 735–742. [CrossRef]
43. Horrocks, N.P.C.; Hegemann, A.; Matson, K.D.; Hine, K.; Jaquier, S.; Shobrak, M.; Williams, J.B.; Tinbergen, J.M.; Tieleman, B.I. Immune Indexes of Larks from Desert and Temperate Regions Show Weak Associations with Life History but Stronger Links to Environmental Variation in Microbial Abundance. *Physiol. Biochem. Zool.* **2012**, *85*, 504–515. [CrossRef]
44. Jain, S.; Gautam, V.; Naseem, S. Acute-phase proteins: As diagnostic tool. *J. Pharm. Bioallied Sci.* **2011**, *3*, 118–127. [CrossRef]

45. Richter, D.; Matuschka, F.-R.; Spielman, A.; Mahadevan, L. How ticks get under your skin: Insertion mechanics of the feeding apparatus of Ixodes ricinus ticks. *Proc. R. Soc. B Biol. Sci.* **2013**, *280*. [CrossRef] [PubMed]
46. Sudhindra, P.; Wang, G.; Schriefer, M.E.; McKenna, D.; Zhuge, J.; Krause, P.J.; Marques, A.R.; Wormser, G.P. Insights into Borrelia miyamotoi infection from an untreated case demonstrating relapsing fever, monocytosis and a positive C6 Lyme serology. *Diagn. Microbiol. Infect. Dis.* **2016**, *86*, 93–96. [CrossRef] [PubMed]
47. Frank, C. Über die Bedeutung von *Laelaps agilis* C.L. Koch 1836 (Mesostigmata: Parasitiformae) für die Übertragung von *Hepatozoon sylvatici* Coles 1914 (Sporozoa: Haemogregarinidae). *Z. Parasitenkd.* **1977**, *53*, 307–310. [CrossRef] [PubMed]
48. Laakkonen, J.; Sukura, A.; Oksanen, A.; Henttonen, H.; Soveri, T. Haemogregarines of the genus *Hepatozoon* (Apicomplexa: Adeleina) in rodents from northern Europe. *Folia Parasitol.* **2001**, *48*, 263–267. [CrossRef]
49. Smith, T.G. The Genus Hepatozoon (Apicomplexa: Adeleina). *J. Parasitol.* **1996**, *82*, 565–585. [CrossRef]
50. Shutler, D.; Smith, T.G.; Robinson, S.R. Relationships between leukocytes and *Hepatozoon* spp. In green frogs, *Rana clamitans*. *J. Wildl. Dis.* **2009**, *45*, 67–72. [CrossRef]
51. Davis, A.K.; Maney, D.L.; Maerz, J.C. The use of leukocyte profiles to measure stress in vertebrates: A review for ecologists. *Funct. Ecol.* **2008**, *22*, 760–772. [CrossRef]
52. Reynolds, L.A.; Filbey, K.J.; Maizels, R.M. Immunity to the model intestinal helminth parasite Heligmosomoides polygyrus. *Semin. Immunopathol.* **2012**, *34*, 829–846. [CrossRef]
53. Martin, L.B.; Burgan, S.C.; Adelman, J.S.; Gervasi, S.S. Host Competence: An Organismal Trait to Integrate Immunology and Epidemiology. *Integr. Comp. Biol.* **2016**, *56*, 1225–1237. [CrossRef]

Article

Measuring Faecal Glucocorticoid Metabolites to Assess Adrenocortical Activity in Reindeer

Şeyda Özkan Gülzari [1,†], Grete Helen Meisfjord Jørgensen [1,*], Svein Morten Eilertsen [1], Inger Hansen [1], Snorre Bekkevold Hagen [1], Ida Fløystad [1] and Rupert Palme [2]

1 Norwegian Institute of Bioeconomy Research, P.O. Box 115, 1431 Ås, Norway; seyda.ozkan@wur.nl (Ş.Ö.G.); svein.eilertsen@nibio.no (S.M.E.); inger.hansen@nibio.no (I.H.); snorre.hagen@nibio.no (S.B.H.); ida.floystad@nibio.no (I.F.)

2 Department of Biomedical Sciences, University of Veterinary Medicine, Veterinärplatz 1, A-1210 Vienna, Austria; rupert.palme@vetmeduni.ac.at

* Correspondence: grete.jorgensen@nibio.no

† Present address: Wageningen Livestock Research, Wageningen University & Research P.O. Box 338, 6700 AH Wageningen, The Netherlands.

Received: 2 October 2019; Accepted: 13 November 2019; Published: 18 November 2019

Simple Summary: We validated an 11-oxoaetiocholanolone enzyme immunoassay for measuring faecal cortisol metabolites (FCMs) in reindeer. Samples were collected from eight male reindeer following adrenocorticotrophic hormone (ACTH) stimulation and from another group of reindeer during handling and calf marking. The overall FCM levels peaked after seven to eight hours in both locations, proving that the assay is suited to evaluate the adrenocortical activity in reindeer.

Abstract: Several non-invasive methods for assessing stress responses have been developed and validated for many animal species. Due to species-specific differences in metabolism and excretion of stress hormones, methods should be validated for each species. The aim of this study was to conduct a physiological validation of an 11-oxoaetiocholanolone enzyme immunoassay (EIA) for measuring faecal cortisol metabolites (FCMs) in male reindeer by administration of adrenocorticotrophic hormone (ACTH; intramuscular, 0.25 mg per animal). A total of 317 samples were collected from eight male reindeer over a 44 h period at Tverrvatnet in Norway in mid-winter. In addition, 114 samples were collected from a group of reindeer during normal handling and calf marking at Stjernevatn in Norway. Following ACTH injection, FCM levels (median and range) were 568 (268–2415) ng/g after two hours, 2718 (414–8550) ng/g after seven hours and 918 (500–6931) ng/g after 24 h. Levels were significantly higher from seven hours onwards compared to earlier hours ($p < 0.001$). The FCM levels at Stjernevatn were significantly ($p < 0.001$) different before (samples collected zero to two hours; median: 479 ng/g) and after calf marking (eight to ten hours; median: 1469 ng/g). Identification of the faecal samples belonging to individual animals was conducted using DNA analysis across time. This study reports a successful validation of a non-invasive technique for measuring stress in reindeer, which can be applied in future studies in the fields of biology, ethology, ecology, animal conservation and welfare.

Keywords: reindeer; stress; glucocorticoids; validation

1. Introduction

The source of stress experienced by animals can be grouped into three categories: (i) physical due to disease or injury; (ii) physiological, for example, due to hunger or temperature control; and (iii) behavioural or psychological, for example, due to a change in living environment. Regardless of the cause, mammals exposed to stress are equipped with defense mechanisms, whereby catecholamines and glucocorticoids (GCs) are released from the adrenals [1]. In spite of their main task to eliminate the effects of the stressor, GCs, when secreted at high amounts over a prolonged time, may pose a risk to animal welfare. Therefore, measurement of cortisol as an indicator of stress is a useful approach in determining changes in health and biophysical parameters, even before the symptoms occur [2,3].

However, cortisol, like many other hormones, is secreted in an episodic and diurnal pattern in many animal species, making it difficult to draw conclusions from single samples. Further, handling animals during the collection of blood samples exposes them to additional stress and may produce results that may reflect the consequences of handling procedures. Therefore, non-invasive and "feedback-free" sampling methods, such as collection of faeces, are preferred [2,4]. Concentrations of faecal cortisol metabolites (FCMs) reflect secreted GCs better than plasma cortisol concentrations estimated at the time of blood collection, providing a reliable approach for measuring adrenocortical function [5,6].

Reindeer (*Rangifer Tarandus Tarandus L.*) are free-ranging, semi-domesticated animals that are not accustomed to being handled by humans in the same way as cattle, sheep or pigs [7]. Gathering, sorting and handling may, therefore, result in increased levels of blood cortisol and urea concentrations [8]. Stress results in increased pH levels and reduced glycogen reserves in the meat, compromising the meat quality [9] and, hence, the shelf life of reindeer products [10,11].

A physiological validation of a non-invasive method aims at specifically stimulating the adrenal glands to release GCs into circulation, which should be well captured in the excreted faecal metabolites [12]. Such validations also yield information about the species-specific time delay between ACTH stimuli and detection of increased FCM levels—a time delay that also corresponds roughly to gut passage time [5]. Species differences in metabolism and excretion of glucocorticoids require that each method is validated for each species and sex [12]. The determination of faecal stress hormone metabolites in mammals and birds has been performed in many species [5,13]. However, it appears that only Ashley et al. [14] have previously performed an ACTH test in reindeer. Since their study did not show an expressed response in FCMs even after a high dose of ACTH, a full validation of such a method in reindeer still appears to be lacking. Thus, this paper aims to evaluate the physiological relevance of an enzyme immunoassay (EIA) for determining FCMs in reindeer through an ACTH challenge, in addition to assessing FCM concentrations of reindeer during calf marking as a biological validation.

2. Materials and Methods

In order to obtain enough reference data to validate the FCMs, two separate studies were performed. The ACTH challenge test was done on eight animals in a controlled fence facility at Tverrvatnet and collection of faecal samples for biological validation was performed for a whole flock of reindeer, representing different ages and both sexes, at Stjernevatn.

2.1. Reindeer (Both Regions) and ACTH Challenge Test (Only Tverrvatnet)

At Tverrvatnet, eight reindeer males (≥1.5 years of age) were injected intramuscularly with 1 mL adrenocorticotrophic hormone (ACTH; 0.25 mg/mL: Synacthen®; CD Pharma Srl & CD Pharmaceuticals AB, Sweden) in the neck area in front of the shoulder. After the injection of ACTH, animals were moved to a designated fenced area where they were not disturbed except for the collection of samples or refilling of food. At Stjernevatn, on the other hand, the animals were gathered for calf marking. The herd consisted of females (lactating and non-lactating), males, yearlings and calves.

2.2. Study Area and Field Conditions

The physiological validation study was carried out at Tverrvatnet (66° N) in Rana municipality in Nordland County in Norway in January 2018. The area was covered with snow and the ambient temperature ranged from −10 to −15 °C. In addition, a biological validation was performed at Stjernevatn (70° N) in Tana municipality in Finnmark County in Norway in August 2017. The ambient temperature at Stjernevatn ranged from 5 to 13 °C.

2.3. Faecal Pellet Collection

In both locations, faecal samples were collected opportunistically, resulting in a challenge to quickly discriminate the samples belonging to adults or calves. Flasko et al. [15] suggest that pellet size could be used as a significant criterion to distinguish calves from adults in wild Canadian caribou (*Rangifer tarandus caribou*). Therefore, to overcome the problem, we allocated the samples with an apparent small pellet size to calves.

2.3.1. Tverrvatnet

To avoid confounding effects of separation stress, the experimental animals at Tverrvatnet were not moved to the experimental enclosure before the injection of ACTH. Faecal samples were collected from the group instead of from each individual reindeer to avoid disturbing the animals and creating additional stress. The identification of individual animals was difficult due to the limited daylight in this area in winter. Since the DNA analysis conducted later revealed the individual identity of the anonymous samples, the above-mentioned challenges were not seen as an obstacle.

The collection of samples started two hours after the injection of ACTH and was repeated every hour until the 12th h and every second hour until 30 h and finished by taking a final sample at 44 h. All samples (n = 317) were collected in plastic bags or in disposable gloves and frozen at ambient temperature (between −10 and −20 °C). Samples were kept frozen at −18 °C until extraction in the laboratory.

2.3.2. Stjernevatn

The sample (n = 114) collection took place from 10:30 to 20:35, forming a total of around 10 h, during two phases: (i) separation of males and non-lactating females from lactating females and calves; and (ii) earmarking of calves. In the first phase, a subgroup of 12 to 30 reindeer were moved into the handling fence where animals were caught by hand and examined by the reindeer herders. The females, which were lactating, were marked with a colour spray and later released back into the subgroup where they eventually remained together with all calves. The males or non-lactating females were moved to adjacent holding pens for separation. Each batch took around 2–8 min to process, depending on the batch size. This phase took approximately five hours, during which a total of 800–900 animals were processed. Faecal samples collected during the first two hours of the separation process were considered as baseline (n = 51, of which n = 31 from known calves).

The second phase of faeces collection was done when calves and lactating females were brought back to the fence for earmarking. In the same way as before, smaller subgroups were taken into the work fence in batches and processed there. Between batches, researchers collected fresh faecal samples from the ground (n = 63, of which n = 34 from known calves) to determine potential changes in FCM concentrations as an effect of handling. This phase lasted 2.5 h. At Stjernevatn, faecal samples could not be attributed to individual reindeer, but stools with smaller pellets were marked as belonging to calves, as described above.

2.4. Analysis of FCMs

Some of the samples from Tverrvatnet were already covered in snow and, to avoid the extra weight snow adds, all samples were oven-dried at 75 °C for 24 h. The dried samples were ground and homogenized using a mortar and pestle, and 0.2 g of each sample was weighed. Steroid extraction was performed according to Palme et al. [16] and Palme [17]. For suspension of samples, 4 mL 99.9% methanol was mixed with 1 mL distilled water and added to the 0.2 g faecal sample. The mixture was hand-vortexed for 1–2 min followed by centrifugation at 2.500 g for 15 min. A 0.5 mL aliquot of each supernatant was transferred to microtubes and stored at −18 °C until analysis.

FCMs were measured in aliquots (after further 1:10 dilution with assay buffer) of the extracts, utilizing a group-specific enzyme immunoassay (11-oxoaetiocholanolone EIA), previously described in detail by Möstl et al. [18], which was found to be well suited in various ruminant species [5]. Intra- and inter-coefficients of variation were below 10 and 15%, respectively.

2.5. DNA Analysis

To identify the individual in the ACTH challenge test that the faecal samples belonged to, we performed DNA analysis on 303 of the 317 samples from Tverrvatnet. A total of 14 samples were discarded at this point, due to low quality.

2.5.1. DNA Extraction

One faecal pellet from each of the 303 faecal samples was transferred to a stool collection tube containing 8 mL stool DNA stabilizer. The DNA was then extracted using PSP Spin Stool DNA Plus Kit (Stratec, Birkenfeld, Germany) following the manufacturers protocol.

2.5.2. PCR Amplification

To reliably assign samples to individual reindeers, we used eight dinucleotide microsatellite markers that were sorted into two multiplex assays for efficient genotyping (multiplex 1 and 2); six of the markers were taken from Wilson et al. [19] (RT1, RT6, RT13, RT20, RT27, RT30) and two from Røed and Midthjell [20] (NVHRT22, NVHRT46). The markers can be found in GeneBank using the accession numbers found in Table 1. The forward-primers were labeled with one of three fluorescent dyes (6FAM, NED or PET, Table 1). In addition, the 'pigtale' sequence [21] was added to the 5′ end of one of the reverse primers to facilitate accurate genotyping.

The PCR reactions were carried out in 10 μL reaction volume: 5 μL 2× multiplex PCR master mix (Qiagen Multiplex kit), 0.05 μg/μL BSA (NEB) and adjusted primer set concentrations (Table 1). The PCR conditions for both multiplexes were 10 min at 95 °C, 35 cycles of 30 s at 94 °C, 30 s at 58 °C, 1 min 72 °C and a final extension for 45 min at 72 °C.

The PCR products (1 μL) were then mixed with Genescan 500 LIZ (Applied Biosystems) size standard (0.24 μL) and Hi-Di formamide (10.00 μL), following a 2 min denaturation at 95 °C on a 2720 Thermal cycler. Capillary electrophoresis was carried out on an ABI 3730 DNA Analyzer (Applied Biosystems). The POP-7™ Polymer was used as a separation matrix and the sample injection times were set to 4 s/2 kv. The PCR fragments were analyzed in GeneMapper 4.1 (Applied Biosystems). The alleles were automatically scored and then manually checked.

To check for possible contamination, every eighth sample added was a negative control. Negative controls contained all of the PCR master-mix components except the DNA template (water was added instead of DNA). Four samples (two tissue- and two scat samples) from previously known reindeer were added as positive controls. Two samples were excluded from the analysis due to an error in registration, leaving 303 samples for the genetic analysis. The combination of the eight markers comprised the genetic profile of an individual reindeer. The samples with the same genetic profile were grouped together and linked to the same individual reindeer.

Table 1. Primer characteristics and PCR conditions for the eight microsatellite markers in *Rangifer tarandus*.

Multiplex Assays	Microsatellite Marker	Primer Sequences (5′–3′)	Repeat Motif	PCR Concentration and Fluorescent Dye	Source	GeneBank Accession No.
1	NVHRT22	F: GTATTCTTGCCAGGAAAAACC R: GTTGCTTCAGTGCTCTCAGAT	CA	0.1 µM, NED	Røed and Midthjell [20]	AF068208
	RT1	F: TGCCTTCTTTCATCCAACAA R: CATCTTCCCATCCTCTTTAC	GT	0.2 µM, FAM	Wilson et al. [19]	U90737
	RT20	F: GCAGAACAGTGAGTGTGAGT R: GTTTCTTGTTGTATTTTGGACCTTT	GT	0.2 µM, NED	Wilson et al. [19]	U90744
	RT13	F: GCCCAGTGTTAGGAAAGAAG R: CATCCCAGAACAGGAGTGAG	GT	0.2 µM, FAM	Wilson et al. [19]	U90743
2	NVHRT46	F: CCGACTGAAGTGACCAAG R: TGTTGAGAGGATTGATAAG	CA	0.2 µM, FAM	Røed and Midthjell [20]	AF068213
	RT6	F: TTCCTCTTACTCATTCTTGG R: CGGATTTTGAGACTGTTAC	GT	0.4 µM, NED	Wilson et al. [19]	U90739
	RT30	F: CACTTGGCTTTTGGACTTA R: CTGGTGTATGTATGCACACT	GT	0.3 µM, NED	Wilson et al. [19]	U90749
	RT27	F: CCAAAGACCCAACAGATG R: TTGTAACACAGCAAAAGCATT	GT	0.2 µM, PET	Wilson et al. [19]	U90748

"Pigtail" sequence added to the 5′ end of the primer according to Brownstein et al. [21].

2.6. Animal Welfare

This study was conducted in accordance with the regulation for use of animals in experiments, adopted by the Norwegian Ministry of Agriculture and Food, and approved by the Ethics Commission on Animal Use by the Norwegian Food and Safety Authority, application number (FOTS ID) 12274 on 03.04.2017. It complies with the EU Directive 2010/63/EU on the use of experimental animals, which was incorporated to the EEA Agreement in May 2015.

2.7. Statistical Analysis

Means, medians and standard deviations (SD) were calculated using data analysis functions in Microsoft Office® Excel. Data were investigated for normal distribution and *log* transformed before analysis. The analysis was generated using SAS software, Version 9.4 of the SAS system for Windows version 6.2.92002 [22].

2.7.1. Samples from Tverrvatnet

First, a non-parametric comparison was performed using the npar1way command with "Animal" as a class variable and "FCM" as the response variable. Using transformed data, the effect of "hour since ACTH administration" on FCM concentrations was analyzed using a mixed model of analysis of variance with "Hour" (1–44) and "Animal" (1–8) as class variables. "Animal" was specified as a random effect and degrees of freedom were calculated using the Satterthwaite's approximation. Differences between means were investigated using the LSmeans command with the Tukey–Kramer approximation.

2.7.2. Samples from Stjernevatn

Using transformed data, the effect of handling on FCM levels was investigated using a general linear model of analysis of variance with "Time of day" (morning or afternoon, reflecting the time periods 0–2 h or 8–10 h after handling) and "Age" (adult/calf) as class variables.

3. Results

3.1. Controlled Experiment with ACTH Challenge on Eight Reindeer (Tverrvatnet)

3.1.1. Identification of Individual Animals

Of the 303 faecal samples analyzed, 48 (16%) were negative for all eight microsatelite markers and 255 (84%) were positive for at least one STR-marker. Among the 255 samples, eight unique genetic profiles were found, representing the eight individual reindeer. A total of 28 of the positive samples (11%) could not be assigned to one of the eight unique profiles, due to the lesser quality of the samples and/or the lack of private alleles or unique allele combinations. The 227 samples (89%) were given an identity based on the eight unique genetic profiles that make up the eight reindeer individuals (Figure 1). As a result, 54 of the samples were assigned to individual 1; 32 samples to individual 2; 15 samples to individual 3; 39 samples to individual 4; 17 samples to individual 5; 16 samples to individual 6; 34 samples to individual 7; and 20 samples to individual 8.

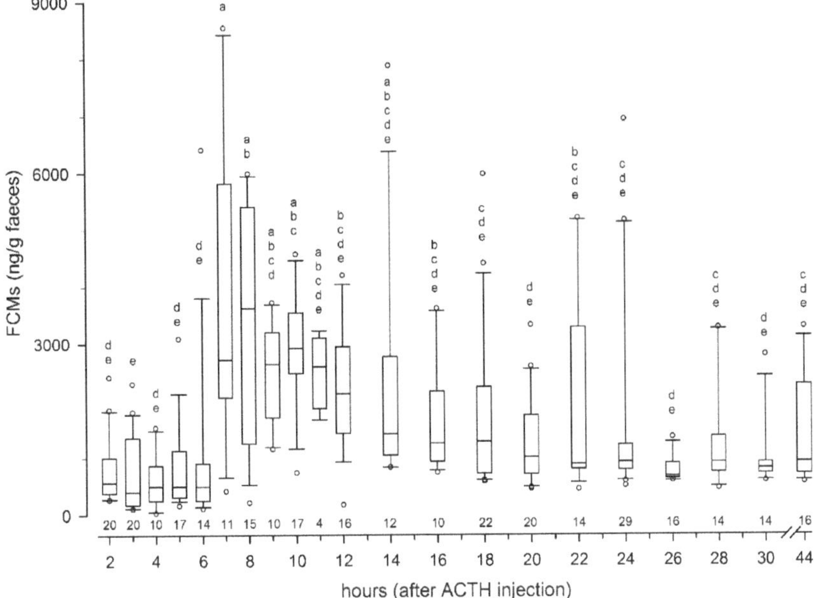

Figure 1. Box-plot graphs of levels of faecal cortisol metabolites of all samples as a function of hours after ACTH administration. Hours with dissimilar letters had significantly different ($p < 0.05$) concentrations. Above the x-axis, the respective numbers of samples are given.

3.1.2. Results from the ACTH Challenge Test

Median (range) FCM levels from two to six hours after injection of ACTH were 505 (34–6408 ng/g). Overall, FCM concentrations peaked at 7 h and slowly decreased afterwards (including high values in some intervals; Figure 1).

When expressed at individual animal levels, the FCM concentrations align with the herd level for five animals where peak levels ranged from 5000 ng to 8000 ng/g faeces. As can be seen in Figure 2, animal numbers 2, 4 and 5 reached their peak values later than the other individuals. There were three unidentified high concentration (6408, 5409 and 3721 ng/g faeces) samples (6–9 h), which could have been those animals' (e.g., animal 5) early peak samples. Individual peak samples were about 6.3 (2.0–16.8) times higher (median 3359; range 2352–5361 ng/g) than respective baselines (0–6 h; 685; 315–1079 ng/g faeces).

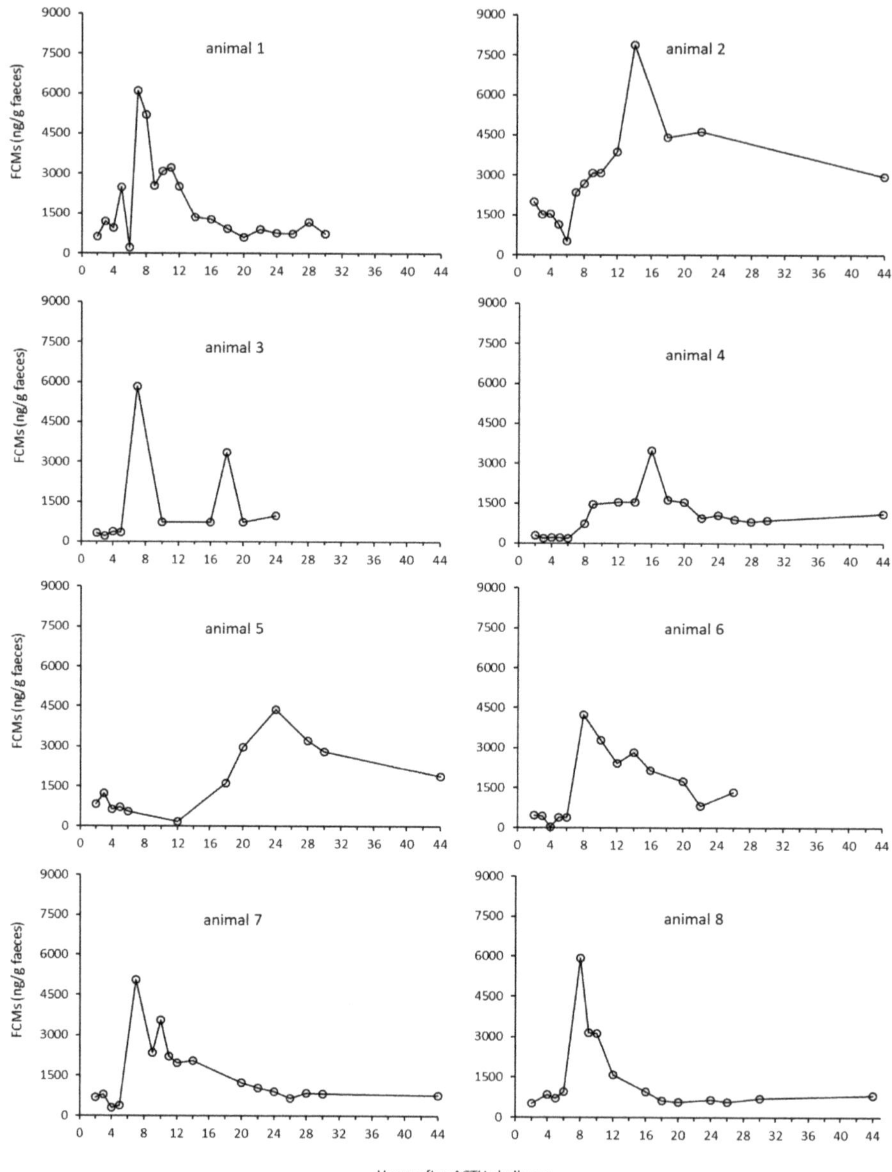

Figure 2. Concentration of faecal cortisol metabolites (FCMs) from identified samples of the eight individual animals at Tverrvatnet, shown as a function of hours after ACTH challenge.

3.2. *Results from the Biological Validation (Normal Handling and Calf Marking at Stjernevatn)*

The highest values of cortisol metabolites were found around eight hours after the human activity started within the fences. FCM concentrations (median; range) in reindeer (Figure 3) were significantly different between morning (480; 212–1159 ng/g faeces) and evening (1469; 605–4673 ng/g faeces) sampling ($F = 116.9$; $p < 0.001$). Calves and adults did not show any significant differences in FCM levels ($F = 1.35$; $p = 0.25$). A total of 65 samples were from calves and 49 samples were from adult reindeer. In the first phase (0–2 h) adult samples were from males and females, non-lactating and lactating. In the second phase (8–10 h) samples from adult reindeer came only from lactating females with calves at foot.

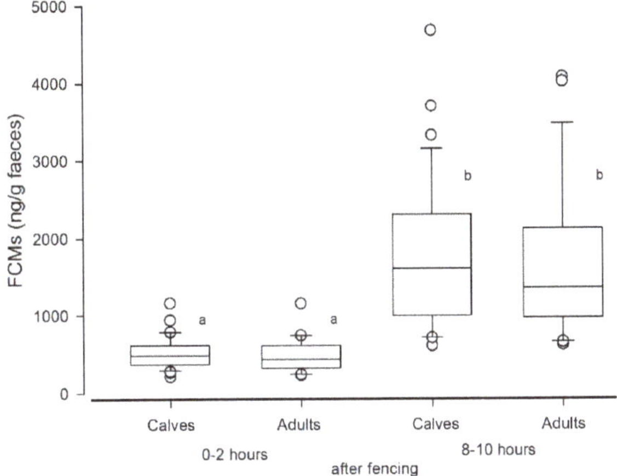

Figure 3. Concentrations of faecal cortisol metabolites (FCMs) in reindeer collected immediately after fencing (0–2 h; morning, n = 51 where 31 were calves and 20 adults) and later after handling and calf marking (8–10 h; evening, n = 63 where 34 were calves and 29 adults). Dissimilar letters indicate significant ($p < 0.001$) differences.

4. Discussion

4.1. Non-Invasive Versus Non-Disturbing

Even though faeces collection is a non-invasive approach compared to blood collection, it may not be non-disturbing, because sample collection can trigger a stress response in animals that are not used to close human presence [4]. This is especially true with herd animals such as reindeer. Since stress is conducive to changes in welfare, production, physiology, health and mortality [23], even slow but repetitive entries into the fence, prompting sudden movements or alertness of the animals, may increase cortisol levels. This will not affect FCM levels in the sample collected immediately but may very well result in higher levels in serial samples collected later on [5]. Thus, the experimental animals receiving ACTH in Tverrvatnet may have already had higher cortisol levels, due to the earlier handling by the reindeer herder moving the animals in the fence. The delayed decrease in FCM levels may be attributed to such a disturbance, because, during the first phase, samples were taken from the herd every hour.

Even though the sample collection at Stjernevatn started once the animals were moved to the working fence, the gathering earlier in the morning may also have influenced or even exacerbated the stress levels of the animals. However, since the time lag between the time the work started in the working fence and the gathering was less than seven hours, the 0–2 h samples most likely reflect true FCM baseline levels.

Sampling protocols, based on earlier experience in similar studies, were prepared in detail before the data collection started. A total number of 317 samples were collected over 44 h in the ACTH challenge test, which should capture any peak levels in FCMs. Earlier studies have demonstrated a time lag of FCM in reindeer of 8 to 24 h [14]. In the biological validation, a total number of 114 samples were collected over 10 h, which could be too short to detect late peak levels. After handling was completed, it was usual practice to release the animals back to pasture. Holding them much longer in fence would, therefore, not be justified in the private herd visited.

4.2. Peak FCM Levels

Following the ACTH challenge, all animals showed well expressed peak FCM levels which were about 630% above their respective baseline levels. This is the first time such a non-invasive method to measure FCM levels in reindeer has been successfully validated. Comparing absolute values in the present study with earlier studies not using the same immunoassays is not relevant, as the different antibodies pick up different cortisol metabolites and to a different degree [5]. The FCM levels peaked seven to eight hours after stimulating the adrenals. This was quicker than reported delay times for deer or goats [24,25], but closer to those found in other ruminants (10–12 h), such as sheep or cattle [6]. However, our results correspond well with findings in caribou, where a significant (but not comparably expressed) increase in FCM was found eight hours after an ACTH challenge [14].

In the biological validation experiment performed at Stjernevatn, the FCM levels eight to ten hours after handling showed more than a three-fold rise compared to baseline levels (samples collected after 0–2 h). This indicates that our method is sensitive enough to detect the effect of handling stress. However, the peak FCM levels were lower when compared to the ACTH challenge test. A study by Carlsson et al. [26] found no increase in FCM levels eight hours after handling stress, but their method was not validated for this species. This underlines the importance of performing validation studies in every animal species and validating each method of FCM analysis against baseline values found under practical and representative conditions.

4.3. Effects of Sex and Individual

At Tverrvatnet, only males were used, because the females were pregnant. Ashley et al. [14] found that female reindeer showed a prolonged response to a very high ACTH dose (8 IU/kg; possibly about 15–16 times higher than the one used here (about 25 IU/animal), assuming the animals weighed about 45–50 kg, whereas males did not have this response. Faecal samples were collected from both males and females in the first collection period at Stjernevatn. In the second period, only lactating females and calves were present. Sex and age may have an influence on FCM levels [5,12]. However, the variation in our dataset from the biological validation was small, as shown in Figure 3. Furthermore, the statistical analysis showed no difference in FCMs between calves and adults, thus making such an influence less likely.

There were quite large individual differences in the FCM response for the ACTH challenge test at Tverrvatnet. Some animals (animals 1, 3, 6, 7, 8) had an early peak (about seven hours after ACTH injection), in others (animals 2, 4 and 5) it occurred later (after 12, 14 and up 24 h, respectively). The reasons for this might be manifold. Differences in the resorption of the ACTH could have played a role. If the needle hits the muscle close to a blood vessel, the response should be quicker (immediate) than if the ACTH injection was further away from a vessel. In case of the latter, the increase could be delayed. Another explanation might be that earlier peaks were missed in those animals. This is likely true in animals (e.g., animal 5) where samples were probably missed, especially as high concentration samples of unknown identity were present in that time window (7–8 h). In addition, the enterohepatic recirculation, which has been proven in ruminants [27], could have caused secondary peaks, especially when they were smaller (e.g., animals 1 and 3). However, median delay times in FCM peaks found in the biological validation at Stjernevatn were also seven hours, which suggests that shorter delay times in faecal peak excretion are more likely found in reindeer.

4.4. Identification of Unknown Samples

Being able to document the individual responses by using DNA analysis added extra value to the results because these individual patterns would have been hidden in anonymous sampling. Non-invasive genetic sampling methods have become widely used in genetic studies and have been shown to be a reliable approach for genetic studies in reindeer [28–30]. It has also been shown that combining genetic analysis with FCM measurements is a useful approach to document individuals and their sex, and that ignoring this information could lead to erroneous conclusions when addressing stress levels in animal populations [31,32].

In this study, we have used non-invasive genetic sampling without any previous knowledge of the genotypes of the eight individuals involved in the experiment and achieved a high success rate in assigning samples to genotypes/individuals across the duration of the study. This shows that the technique can be a useful addition to other scientific fields, to ensure reliable tracking of individuals across time and reduced degree of disturbance (i.e., less human-induced stress responses in the animals). Since we already knew the sex of the eight reindeer individuals, a molecular sexing was not performed, but is advised in future studies without this knowledge, whenever sex can be expected to be an important explanatory factor.

Samples collected by non-invasive methods, such as faecal samples, are less likely to give complete profiles due to low quality and/or quantity (i.e., allelic dropout or fragmented DNA) compared to blood and tissue samples [33–35]. When possible, we therefore recommend including control samples (tissue or blood) from all individuals, collected either before or after the experiment, to confirm their genetic profiles. This would facilitate unambiguous assignment of samples to genetic profiles/individuals and thus contribute to a higher percentage of the samples being available for the primary analysis, e.g., FCM or other applications.

4.5. Ways Forward

Using an animal as its own control in the statistical analysis gave us the opportunity to use fewer animals in the ACTH challenge test. This is in line with the three R's from the ethical code of conduct for animal experiments.

Reindeer are free-ranging ungulates that move over large distances during the production year. Factors including natural and climatic changes affect their welfare. FCM may be used as an objective measure of stress load under different handling procedures, or when human activity interferes with grazing areas and reindeer habitats. The non-invasive faecal sample collection from the ground is of great benefit, when working with this fearful and flock-dependent species. Another benefit is the fact that we have discovered a time-lag of around seven to eight hours between the moment a stressor is experienced until peak FCM levels are reached. Knowing this is imperative when investigating the effects of known potential stressors and comparing them to FCM levels collected before, during and after periods of interference.

5. Conclusions

We have successfully validated an 11-oxoaetiocholanolone EIA for measuring FCMs in order to evaluate adrenocortical activity in reindeer. We found a time-lag of approximately seven to eight hours between the ACTH challenge and peak FCM levels. A similar time-lag was observed when samples were taken after human handling and calf marking in the fence. The method proved biologically sensitive and could be of great value as a tool for welfare assessment of different treatments and handling procedures, and also for evaluating the negative effects of human disturbance in natural pasture areas. As collecting individual faecal samples in a group of housed, semi-free species like reindeer may disturb the animals, combining FCM analysis with genetic data proved useful to overcome the problem of anonymous sampling.

Author Contributions: Conceptualization, G.H.M.J., S.M.E., R.P. and Ş.Ö.G.; Formal analysis, G.H.M.J. and R.P.; Investigation, Ş.Ö.G., G.H.M.J., R.P., I.H., S.M.E., S.B.H. and I.F.; Methodology, R.P. and Ş.Ö.G.; Project administration, Ş.Ö.G., I.H., S.M.E., G.H.M.J., R.P., S.B.H. and I.F.; Supervision, R.P. and Ş.Ö.G.; Visualization, Rupert Palme; Writing—original draft, Ş.Ö.G., R.P., G.H.M.J., S.B.H., I.F., I.H. and S.M.E.; Writing—review and editing, Ş.Ö.G., G.H.M.J. and R.P.

Funding: This paper was produced as part of the project "Handling stress in reindeer" funded by grants for development of arctic agriculture, administered by the County Governor of Troms (project number 10035.01).

The project had great synergy effect with the INTERREG Botnia-Atlantica funded project Animal Sense, administered by Nordland County Council.

Acknowledgments: Authors would like to thank Benan Gülzari, Roberts Stüritis, Arne Johan Lukkassen and Norvald Ruderaas for the technical and administrative support provided during the trial and processing of samples at NIBIO Tjøtta and Edith Klobetz-Rassam for EIA analysis. We also thank Cecilie Marie Mejdell and Solveig Stubsjøen for their inputs during the project development and grant application. Our utmost gratitude goes to the Reindeer herder Tom Lifjell at Tverrvatnet for giving us access to his animals, and their help with the practical work during the ACTH challenge test. Reindeer herders Stig Rune Smuk, Frode Utsi and the other members of Rákkonjárga reindeer district 7 welcomed us into their corrals and made it possible for us to gather faecal samples. Finally, we thank the anonymous reviewers for their valuable comments.

Conflicts of Interest: The authors declare no conflicts of interest.

References

1. Axelrod, J.; Reisine, T.D. Stress hormones: Their interaction and regulation. *Science* **1984**, *224*, 452–459. [CrossRef] [PubMed]
2. Sheriff, M.J.; Dantzer, B.; Delehanty, B.; Palme, R.; Boonstra, R. Measuring stress in wildlife: Techniques for quantifying glucocorticoids. *Oecologia* **2011**, *166*, 869–887. [CrossRef]
3. Palme, R. Monitoring stress hormone metabolites as a useful, non-invasive tool for welfare assessment in farm animals. *Anim. Welf.* **2012**, *21*, 331–337. [CrossRef]
4. Möstl, E.; Palme, R. Hormones as indicators of stress. *Domest. Anim. Endocrinol.* **2002**, *23*, 67–74. [CrossRef]
5. Palme, R. Non-invasive measurement of glucocorticoids: Advances and problems. *Physiol. Behav.* **2019**, *199*, 229–243. [CrossRef] [PubMed]
6. Palme, R.; Robia, C.; Messmann, S.; Hofer, J.; Möstl, E. Measurement of faecal cortisol metabolites in ruminants: A non-invasive parameter of adrenocortical function. *Wien. Tierarztl. Monatsschr.* **1999**, *86*, 237–241.
7. Reimers, E.; Miller, F.L.; Eftestøl, S.; Colman, J.E.; Dahle, B. Flight by feral reindeer *Rangifer tarandus tarandus* in response to a directly approaching human on foot or on skis. *Wildl. Biol.* **2006**, *12*, 403–414. [CrossRef]
8. Rehbinder, C. Management stress in reindeer. *Rangifer* **1990**, *10*, 267–288. [CrossRef]
9. Wiklund, E. Pre-Slaughter Handling of Reindeer (*Rangifer tarandus tarandus L.*)-Effects on Meat Quality. Ph.D. Thesis, Swedish University of Agricultural Sciences (SLU), Uppsala, Sweden, 20 December 1996.
10. Gregory, N.G.; Grandin, T. *Animal Welfare and Meat Science*; CABI Pub.: Wallingford, UK, 1998; p. 298.
11. Jørgensen, G.H.M.; Mejdell, C.M.; Stubsjøen, S.M.; Özkan Gülzari, Ş.; Rødbotten, R.; Bårdsen, B.-J.; Rødven, R. Velferdskriterier i reindriften: En studie av velfersindikatorer og mulige kvalitetskriterier for rein og reinkjøtt. In *Welfare Criteria in Reindeer Management: A Study of Welfare Indicators and Possible Quality Criteria for Reindeer and Reindeer Meat*; Norwegian Institute of Bioeconomy Research (NIBIO): Akershus, Norway, 2017; Volume 3, p. 48. (In Norwegian)
12. Touma, C.; Palme, R. Measuring fecal glucocorticoid metabolites in mammals and birds: The importance of validation. *Ann. N. Y. Acad. Sci.* **2005**, *1046*, 54–74. [CrossRef]
13. Palme, R.; Rettenbacher, S.; Touma, C.; El-Bahr, S.; Möstl, E. Stress hormones in mammals and birds: Comparative aspects regarding metabolism, excretion, and noninvasive measurement in fecal samples. *Ann. N. Y. Acad. Sci.* **2005**, *1040*, 162–171. [CrossRef]
14. Ashley, N.; Barboza, P.; Macbeth, B.; Janz, D.; Cattet, M.; Booth, R.; Wasser, S. Glucocorticosteroid concentrations in feces and hair of captive caribou and reindeer following adrenocorticotropic hormone challenge. *Gen. Comp. Endocrinol.* **2011**, *172*, 382–391. [CrossRef] [PubMed]
15. Flasko, A.; Manseau, M.; Mastromonaco, G.; Bradley, M.; Neufeld, L.; Wilson, P. Fecal DNA, hormones, and pellet morphometrics as a noninvasive method to estimate age class: An application to wild populations of Central Mountain and Boreal woodland caribou (*Rangifer tarandus caribou*). *Canad. J. Zool.* **2017**, *95*, 311–321. [CrossRef]
16. Palme, R.; Touma, C.; Arias, N.; Dominchin, M.F.; Lepschy, M. Steroid extraction: Get the best out of faecal samples. *Wien. Tierarztl. Mon.* **2013**, *100*, 238–246.
17. Palme, R. Measuring fecal steroids: Guidelines for practical application. *Ann. N. Y. Acad. Sci.* **2005**, *1046*, 75–80. [CrossRef] [PubMed]
18. Möstl, E.; Maggs, J.; Schrötter, G.; Besenfelder, U.; Palme, R. Measurement of cortisol metabolites in faeces of ruminants. *Vet. Res. Comm.* **2002**, *26*, 127–139. [CrossRef] [PubMed]

19. Wilson, G.; Strobeck, C.; Wu, L.; Coffin, J. Characterization of microsatellite loci in caribou Rangifer tarandus, and their use in other artiodactyls. *Mol. Ecol.* **1997**, *6*, 697–699. [CrossRef]
20. Røed, K.; Midthjell, L. Microsatellites in reindeer, *Rangifer tarandus*, and their use in other cervids. *Mol. Ecol.* **1998**, *7*, 1773–1776. [CrossRef]
21. Brownstein, M.J.; Carpten, J.D.; Smith, J.R. Modulation of non-templated nucleotide addition by Taq DNA polymerase: Primer modifications that facilitate genotyping. *Biotechniques* **1996**, *20*, 1004–1010. [CrossRef]
22. Statistical Analysis System Institute Inc. *SAS/STAT 9.3 for User's Guide*; SAS Institute: Cary, NC, USA, 2011.
23. Moberg, G.P.; Mench, J.A. *The Biology of Animal Stress: Basic Principles and Implications for Animal Welfare*; CABI Pub.: Wallingford, UK, 2000; p. 377.
24. Huber, S.; Palme, R.; Zenker, W.; Möstl, E. Non-invasive monitoring of the adrenocortical response in red deer. *J. Wild. Manag.* **2003**, *67*, 258–266. [CrossRef]
25. Kleinsasser, C.; Graml, C.; Klobetz-Rassam, E.; Barth, K.; Waiblinger, S.; Palme, R. Physiological validation of a non-invasive method for measuring adrenocortical activity in goats. *Wien. Tierärztl. Mon.* **2010**, *97*, 259–262.
26. Carlsson, A.; Mastromonaco, G.; Vandervalk, E.; Kutz, S. Parasites, stress and reindeer: Infection with abomasal nematodes is not associated with elevated glucocorticoid levels in hair or faeces. *Conserv. Physiol.* **2016**, *4*, cow058. [CrossRef] [PubMed]
27. Palme, R.; Fischer, P.; Schildorfer, H.; Ismail, M.N. Excretion of infused ^{14}C-steroid hormones via faeces and urine in domestic livestock. *Anim. Reprod. Sci.* **1996**, *43*, 43–63. [CrossRef]
28. Ball, M.C.; Finnegan, L.; Manseau, M.; Wilson, P. Integrating multiple analytical approaches to spatially delineate and characterize genetic population structure: An application to boreal caribou (Rangifer tarandus caribou) in central Canada. *Conserv. Genet.* **2010**, *11*, 2131–2143. [CrossRef]
29. Klütsch, C.F.; Manseau, M.; Trim, V.; Polfus, J.; Wilson, P.J. The eastern migratory caribou: The role of genetic introgression in ecotype evolution. *Royal. Soc. Open Sci.* **2016**, *3*, 150469. [CrossRef]
30. Jenkins, D.A.; Yannic, G.; Schaefer, J.A.; Conolly, J.; Lecomte, N. Population structure of caribou in an ice-bound archipelago. *Divers. Distrib.* **2018**, *24*, 1092–1108. [CrossRef]
31. Coppes, J.; Kämmerle, J.L.; Willert, M.; Kohnen, A.; Palme, R.; Braunisch, V. The importance of individual heterogeneity for interpreting faecal glucocorticoid metabolite levels in wildlife studies. *J. Appl. Ecol.* **2018**, *55*, 2043–2054. [CrossRef]
32. Rehnus, M.; Palme, R. How genetic data improve the interpretation of results of faecal glucocorticoid metabolite measurements in a free-living population. *PLoS ONE* **2017**, *12*, e0183718. [CrossRef]
33. Taberlet, P.; Luikart, G. Non-invasive genetic sampling and individual identification. *Biol. J. Linn. Soc.* **1999**, *68*, 41–55. [CrossRef]
34. Fernando, P.; Vidya, T.; Rajapakse, C.; Dangolla, A.; Melnick, D. Reliable noninvasive genotyping: Fantasy or reality? *J. Hered.* **2003**, *94*, 115–123. [CrossRef]
35. Ball, M.C.; Pither, R.; Manseau, M.; Clark, J.; Petersen, S.D.; Kingston, S.; Morrill, N.; Wilson, P. Characterization of target nuclear DNA from faeces reduces technical issues associated with the assumptions of low-quality and quantity template. *Conserv. Genet.* **2007**, *8*, 577–586. [CrossRef]

Article

Towards Non-Invasive Methods in Measuring Fish Welfare: The Measurement of Cortisol Concentrations in Fish Skin Mucus as a Biomarker of Habitat Quality

Annaïs Carbajal [1,*], Patricia Soler [2], Oriol Tallo-Parra [3], Marina Isasa [4], Carlos Echevarria [4], Manel Lopez-Bejar [1,†] and Dolors Vinyoles [2,†]

1 Department of Animal Health and Anatomy, Veterinary Faculty, Universitat Autònoma de Barcelona, Bellaterra, 08193 Barcelona, Spain; manel.lopez.bejar@uab.cat
2 Department of Evolutionary Biology, Ecology and Environmental Sciences, Universitat de Barcelona, Avinguda Diagonal 643, 08028 Barcelona, Spain; patrisoler89@gmail.com (P.S.); d.vinyoles@ub.edu (D.V.)
3 Department of Animal and Food Science, Veterinary Faculty, Universitat Autònoma de Barcelona, Bellaterra, 08193 Barcelona, Spain; oriol.tallo@uab.cat
4 Cetaqua, Centro tecnológico del agua, Cornellà de Llobregat, 08940 Barcelona, Spain; marina.isasa@cetaqua.com (M.I.); cechevarriadc@cetaqua.com (C.E.)
* Correspondence: anais.carbajal@uab.cat or annais.carbajal@gmail.com
† M.L.-B. and D.V. were co-principal investigators.

Received: 20 September 2019; Accepted: 6 November 2019; Published: 8 November 2019

Simple Summary: The analysis of circulating cortisol has been by far the most common method used as a means to assess fish stress responses and, thus, animal welfare. To avoid many of the drawbacks inherent to blood sampling, cortisol can be less-invasively detected in fish skin mucus. The measurement of cortisol in skin mucus however, has, to date, only been demonstrated as suitable for farm fish, although its application to free-ranging animals would offer many advantages. The present study was therefore designed to evaluate the applicability of skin mucus cortisol analysis as a potential tool to assess habitat quality. To that end, wild fish residing in environments of different habitat quality were sampled for blood and skin mucus. First, several physiological endpoints typically used as indicators of exposure to pollutants were accurately related to the habitat quality in the Catalan chub (*Squalius laietanus*). Second, cortisol levels in blood were also compared between habitats, and they were successfully correlated to skin mucus cortisol concentrations. Finally, we contrasted the patterns of response of all the endpoints assessed to skin mucus cortisol levels across the sites. The strong linkages detected in this study provide new evidence that the measurement of cortisol in skin mucus could be potentially used as a biomarker of habitat quality in freshwater fish.

Abstract: Cortisol levels in fish skin mucus have shown to be good stress indicators in farm fish exposed to different stressors. Its applicability in free-ranging animals subject to long-term environmental stressors though remains to be explored. The present study was therefore designed to examine whether skin mucus cortisol levels from a wild freshwater fish (Catalan chub, *Squalius laietanus*) are affected by the habitat quality. Several well-established hematological parameters and cortisol concentrations were measured in blood and compared to variations in skin mucus cortisol values across three habitats with different pollution gradient. Fluctuations of cortisol in skin mucus varied across the streams of differing habitat quality, following a similar pattern of response to that detected by the assessment of cortisol levels in blood and the hematological parameters. Furthermore, there was a close relationship between cortisol concentrations in skin mucus and several of the erythrocytic alterations and the relative proportion of neutrophils to lymphocytes. Taken together, results of this study provide the first evidence that skin mucus cortisol levels could be influenced by habitat quality. Although results should be interpreted with caution, because a small sample size was collected in one studied habitat, the measurement of cortisol in skin mucus could be potentially used as a biomarker in freshwater fish.

Keywords: non-invasive; bioindicator; pollution; stress; welfare; constructed wetland; glucocorticoid; urban river

1. Introduction

Throughout their lifetime, wild fish face many challenges of the aquatic environment that can impose considerable stress and reduce their welfare [1]. These challenges can be either natural or have an anthropogenic origin, and, depending on the magnitude and duration, they can cause acute or chronic stress responses [2,3]. Acute stress responses, such as those triggered by a predator attack or certain unpredictable weather conditions, can facilitate survival [4], whereas long-term stressors, like exposure to environmental pollution, are associated with a wide range of maladaptive effects [5] that may, ultimately, lead to loss of biodiversity [6,7]. Accordingly, understanding the causes and effects of environmental disturbances on fish physiology may help developing conservation strategies to enhance restoration and protect freshwater ecosystems [5,8].

An economical and practical option that can give a substantial amount of information about the overall health status of individuals is the peripheral blood test [9,10]. The analysis of red blood cells (RBCs) allows the detection of DNA damage and alterations by the assessment of erythrocytic nuclear abnormalities (ENA), circulating micronuclei (MN), and senescent (SE) and immature (IE) erythrocytes [11,12]. The detection of RBC abnormalities has actually been widely used as an indicator of exposure to genotoxic and mutagenic contaminants [13–15]. In parallel, relative white blood cell (WBC) count can be obtained, which offers a very common measure of stress and innate immune response [16]. In particular, the relative proportion of neutrophils to lymphocytes has been successfully applied as a measure of prolonged pollutant exposure [9,17,18]. Other uses of blood samples in ecotoxicology include the quantification of glucocorticoid (GC) hormones, such as cortisol, to assess the stress response [19,20]. Cortisol is the main GC in teleost fish secreted after the activation of the hypothalamic–pituitary–interrenal (HPI) axis in response to acute and chronic stress [2,21]. Analyses of cortisol levels in blood and, more recently, in whole-body and the surrounding water have been effectively used to monitor environmental stress responses [22–24]. Blood, whole-body, and the surrounding water sampling, however, present clear limitations when being applied in wild population studies. First, blood collection is an invasive technique that the process by itself may provoke further stress and thus it can potentially compromise the animal's welfare. Similarly, whole-body cortisol analysis involves sacrifice of the specimens [22]. And finally, collection of the holding water requires fish restriction in a bucket, which can cause additional stress. Moreover, this technique is difficult to apply in the wild [24]. Fish scales can also accumulate cortisol [25,26]; however, their potential as biomarkers of habitat quality deserves further investigation [27]. Cortisol analysis in fish skin mucus has recently gained considerable attraction, especially because the sampling method is much less invasive compared to the aforementioned techniques [28–30]. Skin mucus cortisol levels have been shown to reflect circulating concentrations in several species of farm fish [26,30–32], but there is yet no evidence of such a relationship in free-ranging species. In addition, this method has, to date, only been applied in strictly controlled environments [30,33], hence its applicability in uncontrolled, natural environments remains to be explored. Therefore, the present study aimed to examine whether skin mucus cortisol concentrations (MCC) from the freshwater fish Catalan chub, *Squalius laietanus* [34], are affected by the habitat quality to further develop non-invasive biomarkers in free-living fish. Catalan chub was chosen, as this species has demonstrated to be a good candidate for freshwater biomonitoring using blood tests [11]. It is well known that understanding changes in cortisol levels is not a simple process, especially when measuring cortisol in wild animals by using alternative samples other than blood [6,9]. Given that the measurement of cortisol in skin mucus is a novel method, other physiological endpoints of the effects of pollution in fish were assessed to better interpret cortisol fluctuations in this matrix. Several hematological parameters (RBC anomalies and altered WBC counts) were measured in parallel, since, as previously mentioned, they have successfully been used as indicators of health condition in the Catalan chub [11], as well as in many other species (reviewed above).

This study was carried out in a populated and industrialized urban river, where efforts are being made to minimize the environmental impacts and recover the aquatic fauna throughout the performance of constructed wetlands. Catalan chub were sampled from a non-impacted upstream site and two downstream polluted sites, located within the constructed wetland system, in order to compare fish residing environments of different habitat quality. Initially designed specifically for wastewater treatment, constructed wetlands are nowadays an important component of urban ecosystems since they play a crucial role in environmental pollution control [35–37]. Constructed wetlands are macrophyte-based systems that remove pollutants through a combination of physical, chemical, and biological processes [36,38]. Wetlands' performances, though, need to be periodically monitored [39]. The described methodologies for wetland monitoring include physical and chemical techniques that provide information about the amount of pollutants present in the water. Nevertheless, these tools do not give insight into how living organisms cope with water contaminants [39]. On this basis, the present study was carried out in a constructed wetland system to highlight the need to apply techniques that provide information about how animals perceive and adapt to their environment.

2. Materials and Methods

2.1. Study Area and Field Sampling

In order to study the influence of the habitat quality on skin mucus cortisol concentrations, individuals were sampled from two sites within a wetland system (Besòs River Park, NE Spain), each of which represents a different stage of biodegradation of water pollutants (P1 and P2), and a reference non-impacted upstream site located outside the wetlands (Figure 1). The reference site was set in a small tributary (Riera d'Avencó), 49.6 km distant from the site P1. The sampling site P1 was placed at the beginning of the constructed wetland, 2.9 km distant from P2, which was located at the end of the overall wetland system and 3.6 km to the river mouth. The Besòs River is an urban river adjacent to the City of Barcelona (Catalonia, NE Spain). During the 1970s and 1980s this river was declared the most polluted river in Europe. Fish populations are slowly recovering, making it easy to find an abundance of differences between nearby sites.

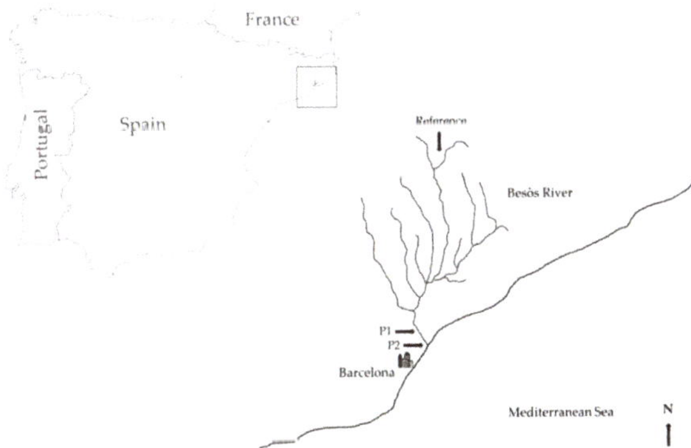

Figure 1. Map showing the location of the three sites sampled in the present study (Reference, P1 and P2) within the Besòs River, in north-east Spain.

Sampling areas were determined following the protocols from the European Committee for Standardization (CEN prEN 14011:2002). Fish were sampled by electrofishing using a portable unit which generated up to 200 V and 3 A pulsed DC in an upstream direction. This capture method was employed as it is considered an easy and safe method, very popular for studying stream fish [2,40,41]. To minimize the potential influence of the sampling technique on the results, fish were all handled identically by the same experienced operator.

At the start of the study (17 May 2017), several physico-chemical parameters and contaminants of emerging concern (CEC) were measured as a basis of the water quality from each sampling site (Table 1). Results of the water analyses provided evidence that the sampling sites classified as polluted presented features typically identified in disturbed environments [11,42,43].

To avoid the circadian rhythm being a source of variability, samplings were performed in the morning (10:00–12:00) from 17 May to 15 June 2017, once the constructed wetland system had been operative for 3 months (23 May, P1; 17 May, P2; 15 June, reference site). Unequal sample size was collected within the polluted sites probably due to sub-optimal habitat conditions in P1 (n = 6) compared to P2 (n = 17) or the reference site (n = 22). All procedures followed the national and institutional regulations of the Spanish Council for Scientific Research (CSIC) and the European Directive 2010/63/EU.

Table 1. Occurrence of contaminants of emerging concern (CEC) and physico-chemical data from river water samples collected within the wetland system (P1 and P2) and the reference non-impacted site located outside the system (Reference) in May 2017, when fish were sampled.

Compound	Sites		
	P1	**P2**	**Reference**
CEC (μg/L)			
Volatile organic compounds			
Tetrachloroethene	<LOD	0.6	<LOD
Pesticides			
Simazine	0.13	0.13	<LOD
Diuron	<LOD	<LOD	<LOD
Isoproturon	0.04	0.04	<LOD
Pharmaceutical products			
Diclofenac	1.61	0.29	<LOD
Alkylphenols			
4-tert-octylphenol	0.025	<LOD	<LOD
Nonylphenol	0.14	<LOD	<LOD
Physico-Chemical Data (mg/L)			
NH_4^+	12.6	10.6	0.07
NO_3^-	3.31	2.83	0.18
PO_4^-	0.8	1	0.4
TOC [1]	9.25	6.39	2.16
COD [2]	29.9	30.5	5.88
SS [3]	7	9.5	0.5
Turbidity (NTU [4])	4.47	3.01	0.66

Concentrations of compounds under the instrumental detection limit (LOD, Limit of detection) are not included. [1] TOC, total organic carbon; [2] COD, chemical oxygen demand; [3] SS, suspended solids; [4] NTU, nephelometer turbidity units.

2.2. Sample Collection

Fish were sampled after applying the combined stressor of capture and a brief period of confinement, given that stress-induced cortisol levels offer a considerable understanding of the overall stress response [44]. In this context, growing evidence suggests that circulating cortisol increases can be detected from as short as 1–2.5 min following exposure to stressors [1,21]. Accordingly, fish were caught using the portable electrofishing unit, and confined in buckets of 20 L for approximately 15 min. Afterwards, specimens were anesthetized with MS-222 (100 mg/L) and immediately after, blood and skin mucus were sampled. A heparinized insulin syringe was used to collect approximately 0.2 mL of blood by caudal vein puncture. A drop of blood was smeared for hematological analyses, and the remaining fluid kept on ice until transported back to the laboratory. Samples were then centrifuged (1500× g, 5 min) and plasma was collected and stored at −20 °C. Skin mucus was collected following the method described by Schultz and colleagues [45] with some modifications. Briefly, a polyurethane sponge was used to absorb the skin mucus by applying light pressure to the left and right flank as this method has been shown to be less stressful than using a spatula [45]. The sponge was then introduced into a cylinder of a syringe and compressed with the barrel to collect, into a centrifuge tube, the skin mucus. Afterwards, samples were centrifuged (2000× g, 10 min) and the supernatant was stored at −20 °C until analysis. Morphological variables, including length (mean ± SD: P1 = 208.5 ± 34.2 mm; P2 = 211.6 ± 22.8 mm; reference = 159.9 ± 35.5 mm) and weight (mean ± SD: P1 = 126.7 ± 61.1 g; P2 = 117.0 ± 50.0 g; reference = 58.7 ± 42.2 g) were measured. Fulton's body condition factor (K), calculated according to the formula K = 100,000 body weight (g) total length (mm)$^{-3}$ [46], was assessed, since it can reflect the energetic state of individuals [47]. Fish were released into the corresponding capture site once the samples had been collected.

2.3. Hematological Analysis

Immediately after being collected, a drop of blood was placed on glass microscope slides, drawn across the surface and, once air-dried, slides were fixed in absolute methanol for 10 min. This procedure was run in duplicate for each specimen. Upon arrival in the laboratory, one of each duplicated slides was stained with Diff-Quick® to assess the frequency of abnormal RBCs and for the WBC count. RBCs (1000) of each individual slide were scored to calculate the frequency of ENA, SE, and IE. The ENAs analyzed were defined as lobed, kidney-shaped, fragmented, and vacuolated nuclei following the directions of Pacheco and Santos [48]. The relative count of all types of WBCs (neutrophils, lymphocytes, monocytes, eosinophils, and basophils) was performed for 100 WBCs following the directions of Tavares-Dias [49]. The neutrophil and lymphocyte count was used to calculate the relative proportion of neutrophils to lymphocytes (hereafter, N:L ratio). The second duplicated slide of each individual was used to assess the number of micronucleus after performing an acridine orange staining. RBCs (3000) of each slide were scored to calculate the frequency of MN.

2.4. Cortisol Extraction and Biochemical Validation

To analyze cortisol levels from blood and skin mucus, a commercial enzyme immunoassay (EIA) kit (Cortisol ELISA KIT; Neogen® Corporation, Ayr, UK) was used.

Biochemical validation of the EIA was carried out following the methods described by Carbajal and collaborators [50]. Samples of plasma and skin mucus extracts from several individuals were first pooled and used in each validation test. Precision was evaluated with the intra-assay coefficient of variation (CV), calculated from all duplicated samples analyzed. All samples from each matrix evaluated were analyzed in single assays; therefore, the inter-assay CV was not assessed. The dilution test was applied to assess the specificity of the EIA kit by comparing observed and theoretical values from pools diluted with EIA buffer. To test the assay's accuracy, the spike-and-recovery test was used, where known volumes of pools were mixed with different volumes and concentrations of pure standard cortisol solution. Finally, we evaluated the sensitivity of the test, given by the smallest amount of cortisol concentration detected.

2.5. Statistical Analysis

The computer program R software (R-project, Version 3.0.1, R Development Core Team, University of Auckland, New Zealand) was used to analyze the data. A $p < 0.05$ was considered statistically significant. Normality of the data was assessed using Shapiro–Wilk tests, and parametric or non-parametric tests were applied accordingly. Differences in cortisol levels and hematological data between sites were assessed using one-way ANOVA with Tukey's pairwise post-hoc tests.

Non-normally distributed data were assessed by using Kruskal–Wallis test, followed by a multiple comparison test. Length and K were run as covariates in the models to account for potential differences across sites due to the age of the fish or the energy accumulated in the body, respectively. These covariates were then removed when results showed no influence on the response variable ($p > 0.05$). Pearson and Spearman correlation tests were applied to test for correlations between skin mucus cortisol levels to levels of the hormone in blood and the hematological variables.

3. Results

In total, 45 Catalan chub from sites P1 (mean $K \pm$ SD = 1.17 ± 0.14, N = 6), P2 (mean $K \pm$ SD = 1.30 ± 0.07, N = 17), and reference (mean $K \pm$ SD = 1.18 ± 0.08, N = 22) were captured, sampled, and returned to the river immediately after sampling.

3.1. Biochemical Validation of the EIA

Plasma and skin mucus intra-assay CV was 8.8% and 7.7%, respectively. The dilution test obtained for plasma showed an R^2 = 98.4% and a mean percentage error of 104.1 ± 4.1%. In the skin mucus dilution test, an R^2 = 99.7% and a mean percentage error of 108.7 ± 8.7% was obtained. Also, in the dilution test, obtained and theoretical concentrations of plasma and skin mucus extracts showed significant correlation coefficients ($r = 0.99$, $p < 0.01$). In the spike-and-recovery test, the average of the recovery percentage was 107.6 ± 10.0% for plasma and 109.6 ± 9.1% for skin mucus. The sensitivity of the assay for plasma and skin mucus assessment was 0.07 ng cortisol/mL and 0.03 ng cortisol/mL, respectively. The biochemical validation of the EIA showed reliable results that demonstrated the assay's precision, specificity, accuracy and sensitivity in measuring plasma and skin mucus cortisol levels of the Catalan chub.

3.2. Hematological Parameters and Cortisol Levels

Concerning the RBC alterations, the frequency of IE (Figure 2A) and SE (Figure 2B) was significantly lower in the reference site compared to P2 ($p < 0.05$). In addition, significantly lower frequencies of ENA (Figure 2C) were detected in the reference site compared to both polluted habitats ($p < 0.05$). Despite no differences detected between polluted and reference sites in the frequency of MN ($p > 0.05$), a higher frequency of MN was detected in P1 compared to P2 ($p < 0.05$).

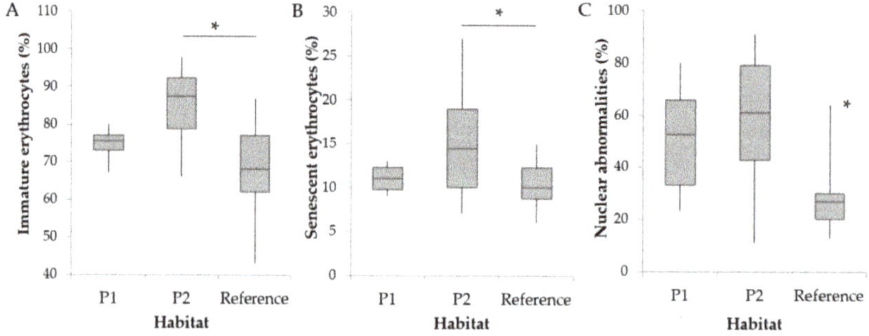

Figure 2. Frequencies of (**A**) immature erythrocytes, (**B**) senescent erythrocytes, and (**C**) erythrocytic nuclear abnormalities determined in the Catalan chub sampled at two polluted sites (P1, n = 6; P2, n = 17) and a reference non-impacted upstream site (n = 22) in the Besòs River. Asterisks (*) indicate significant differences between habitats ($p < 0.05$).

When accounting for differences in WBC counts between habitats (Table 2), we detected a higher N:L ratio in P1 and P2 compared to the reference site ($p < 0.05$), and a significant interaction effect of site with length on the N:L ratio ($p < 0.05$). Although no differences were detected between habitats in the proportion of monocytes and eosinophils ($p > 0.05$), a higher proportion of basophils was detected in P2 compared to the reference site ($p < 0.05$).

Table 2. Mean values and standard deviation of white blood cell parameters (‰) determined in the Catalan chub from polluted (P1, n = 6; P2, n = 17) and reference (n = 22) sites in Besòs River. Different letters indicate significant differences among sites ($p < 0.05$).

White Blood Cell Type	Sites		
	P1	**P2**	**Reference**
N:L ratio	10.54 ± 4.49 [a]	8.00 ± 3.46 [a]	4.79 ± 2.91 [b]
Monocytes	4.50 ± 3.67	7.47 ± 3.83	6.05 ± 2.82
Eosinophils	0.67 ± 0.82	1.35 ± 1.17	1.18 ± 1.30
Basophils	0.33 ± 0.52 [ab]	1.24 ± 1.20 [a]	0.36 ± 0.58 [b]

Plasma cortisol concentrations (PCC) (Figure 3A) and mucus cortisol concentrations (MCC) (Figure 3B) were significantly lower in the reference site than in both polluted sites ($p < 0.05$).

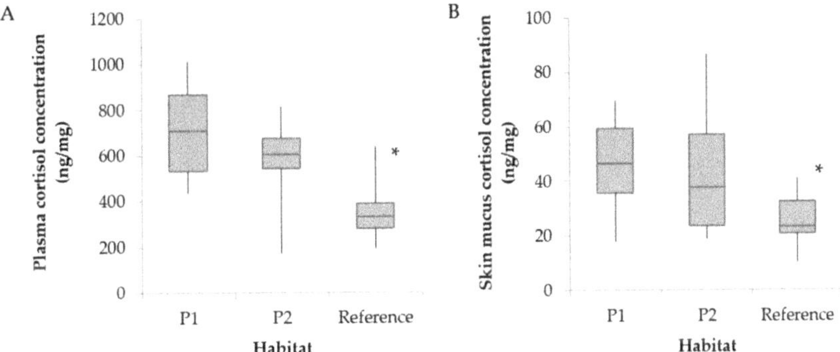

Figure 3. Cortisol concentrations in (**A**) plasma and (**B**) skin mucus of the Catalan chub sampled at two polluted sites (P1, n = 6; P2, n = 17) and a reference non-impacted upstream site (n = 22) in the Besòs River. Asterisks (*) indicate statistically significant differences between habitats ($p < 0.05$).

When studying the relationships between MCC, PCC, and the hematological parameters (Table 3), significant correlations were identified between MCC and PCC ($p < 0.01$), IE ($p < 0.05$), ENA ($p < 0.05$), and the N:L ratio ($p = 0.05$).

Table 3. Correlation (*r*) and *p*-value between hematological variables and skin mucus cortisol concentrations (MCC).

Variable	MCC (*r*)	*p*-Value
Cortisol		
Plasma cortisol concentration	0.55	**<0.01**
Red blood cells		
Immature erythrocytes	0.40	**0.03**
Senescent erythrocytes	0.23	0.24
Erythrocytic nuclear abnormalities	0.41	**0.02**
Micronucleus	−0.001	0.99
White blood cells		
N:L ratio	0.34	**0.05**
Monocytes	−0.01	0.97
Eosinophils	−0.21	0.20
Basophils	0.24	0.17

Bold numbers denote significant correlations between the hematological variables and MCC.

4. Discussion

In this study, we first successfully validated that several physiological endpoints typically used as indicators of exposure to pollutants (abnormal RBCs and altered WBC counts), were accurately related to the habitat quality in the Catalan chub. Cortisol levels in blood were also compared between habitats and they were correlated to skin mucus cortisol concentrations. Finally, we contrasted the patterns of response of all the endpoints assessed to skin mucus cortisol levels across the sites. Lack of an adequate number of samples in one of the polluted sites (P1, n = 6) makes the cross-site comparison difficult. The discussion is therefore focused on differences detected between sites with a larger sample size (P2, n = 17; reference, n = 22). Results highlight the potential of this non-invasive tool to assess habitat quality and the need to combine regular techniques for biomonitoring the wetlands systems' performances with the measurement of cortisol in skin mucus.

4.1. Abnormal RBC Frequencies

There was no consistent pattern in abnormal RBC frequencies when they were compared between the polluted and the reference habitats. As confirmed by the physico-chemical and CEC analysis (Table 1), the site P1, located at the beginning of the wetlands system, presented slightly worse habitat conditions compared to the site P2, placed at the end of the same system. Accordingly, we expected to identify further RBC alterations in P1 than in P2. Nevertheless, relative to the reference site, P1 only exhibited significantly higher frequencies of ENA, while fish from P2 presented greater IE, SE, and ENA frequencies. As mentioned earlier, the sub-optimal conditions of P1 were likely the cause of the small sample size collected in the site, which in turn could have limited the statistical power in detecting potential differences.

Conversely, both polluted P1 and P2 sites showed clearly higher ENA levels than the reference habitat. Particularly in spring, greater frequencies of nuclear abnormalities in the same fish species have already been identified [11]. The ENA test has been demonstrated to be a highly sensitive parameter for pollution assessment [12,51], probably explaining the clear variations detected in these nuclear abnormalities between habitats.

Besides the inconsistencies found between both polluted sites, higher frequencies of RBC disorders detected in the polluted habitat with larger sample size further support the idea that IE, SE, and ENA tests are reliable biomarkers of habitat quality in the Catalan chub [52,53].

Although these commonly noted abnormalities are highly sensitive to pollution, they are not as widely accepted as the use of MN tests [53,54]. The influence of river status on the frequency of MN has been previously studied in the Catalan chub, and, in accordance with our results, no changes were detected between degraded and reference streams [11]. Nevertheless, in our study, the two polluted areas evaluated differed in MN levels between them, with the highest values detected at the end of the wetland system. Although this result could also be given by the different sample size between study sites, it should be noted that a different contaminant profile can also result in mismatch between habitats [13]. For example, exposure to atrazine and ametrine herbicides resulted in increased MN [55]; in contrast, a different herbicide, pendimethalin, showed increased ENA but not MN [56]. Tetrachloroethene, the only CEC analyzed that presented higher concentrations in P2 than P1, is a dry-cleaning compound widely used in the textile industry, known to have toxic effects in fish [57,58]. Although evidence in humans supports the link between the MN formation and this compound [59], to the authors' knowledge, there are no published studies demonstrating this association in fish. Despite this, other compounds not specifically analyzed in this study could also be the consequence of the differing results between polluted habitats.

Taken together, these findings suggest that the detection of RBC disorders can be potentially used to identify low-quality habitats for the Catalan chub, while being an approach that could contribute to a better understanding of the species' health status than the MN test.

4.2. Variations in WBC Counts

Characteristic changes in blood leukocyte counts have been generally linked to the continuous activation of the HPI axis [2,60–62]. Interestingly, prolonged exposure to environmental contaminants can cause neutrophilia and/or lymphopenia in fish [16–18], likewise in other taxa [9,63]. In line with these published reports, different WBC counts were detected between the study sites. Most notably, the N:L ratio was significantly higher in both polluted sites compared to the reference stream. These between-site differences could be a consequence of the sub-optimal environmental conditions in P1 and P1 habitats. However, the significant interaction detected between length (age) of fish and site suggests caution when interpreting results. Although length did not influence any of the other response variables tested, it remains possible that differences in the N:L ratio were partly exacerbated by the smaller size of fish sampled at the reference site. Age can be an important factor in shaping the stress response [2]; nevertheless, further studies are needed to understand the influence that age/length can have in the leukocyte profile. The number of basophils, a cell type still not assessed in this species, was also higher in P2 compared to the reference site. Although the function of this cell type is poorly understood, probably because its occurrence in teleost fish seems to be very rare [49,64], basophils have been related to acute inflammation processes [16]. Besides this, neither the monocyte nor the eosinophil count appeared to differ between sites, similar to earlier findings on silver carp (*Hypophthalmichthys molitrix*) in response to pesticides [65]. The assessment of these cell types is not common in contemporary research, perhaps due to the controversy concerning the effect of stress on eosinophil and monocyte numbers [2,16]. When only WBC data are available, evaluation of these two leukocyte types can help distinguish stress from infectious responses [16], thus further research on monocyte and eosinophil changes is strongly encouraged.

4.3. Changes in Cortisol Levels

Cortisol levels detected in plasma and skin mucus within the same individuals displayed a close linear relationship, suggesting that cortisol diffuses to the skin mucus in proportion to the amount of circulating hormone. Validation of alternative matrices for HPI axis activity assessment should prove that hormone concentrations in these media are proportional to their abundance in the bloodstream [3,66], as the present study demonstrates for the measurement of cortisol in skin mucus. These results, therefore, increase the applicability of the method as a sensitive-individual measure of the HPI axis activity in wild freshwater fish within their natural environment.

In addition, both plasma and skin mucus cortisol levels differed significantly between habitats of different quality, with the highest hormone values observed in the polluted sites assessed. This association between cortisol concentrations and habitat quality suggests that variation in the HPI axis activity is likely to be related to the presence of environmental disturbances. Greater stress responses attributed to the effects of pollutants have been reported in several fish species [22,67,68], as well as in other taxa [69–71]. Nevertheless, it is important to note that chronic exposure to certain aquatic contaminants can also have suppressive effects on the stress axis [72–75]. In this context, investigating the toxic mechanisms underlying variation in the HPI axis alteration will be particularly informative.

4.4. Integrated Assessment

Interpreting cortisol fluctuations in free-living vertebrates is certainly a complex practice, particularly when applying alternative matrices for hormone assessment [6,9]. This is why linking cortisol levels to other endpoints of the stress responses can significantly enhance the current understanding on the ecology of stress [76].

In the present study, fluctuations in skin mucus cortisol levels between habitats paralleled those detected in blood, the traditional matrix used for hormone assessments in fish [77]. Relative to habitat quality, changes of the hormone in skin mucus also coincided with variations in the hematological parameters, except for MN levels. Furthermore, the amount of cortisol in skin mucus was directly proportional to frequencies of abnormal erythrocytes (IE and ENA) and to the well-established stress index N:L ratio. Red blood cells are highly sensitive to landscape disturbances [78] and, more specifically, to environmental pollution [79,80]. Accordingly, the measurement of abnormal erythrocytes has been successfully used to assess the health status of the Catalan chub [11,42] and many other fish species [12,13,17]. In the same context, WBC counts, particularly the N:L ratio, increases in individuals exposed to heavy metals and other contaminants proportional, indeed, to the circulating cortisol levels [16]. Given the very clear effect of pollution on leukocyte and erythrocyte profiles, the strong linkages detected in this study provide new evidence that the measurement of cortisol in skin mucus could be potentially used as a biomarker of habitat quality in freshwater fish residing polluted environments.

The use of a robust sample size is recommended in natural settings where individuals are exposed to different environmental conditions [9,81]. However, capturing a relevant number of individuals in wild conditions may not always be possible, especially in highly degraded habitats such as the one included in the present field experiment. An important limitation of our study design may, therefore, be the small sample size in one of the polluted habitats, which requires results of this site be interpreted with caution.

While physical and chemical techniques are commonly applied for wetland monitoring [39], methodologies that provide information about how animals are influenced by their environment are only occasionally used. The incorporation of biomarkers into the constructed wetlands' management would provide complementary data to the conventional analyses. Hence the demonstrated sensitivity of the methods evaluated in the present study to different pollution gradients could be exploited for biomonitoring the wetlands systems' performances. Indeed, the non-invasive measurement of cortisol in skin mucus would largely improve our understanding about the link between the detected chemical concentrations and the biological effects observed.

Author Contributions: Conceptualization, C.E., M.I., D.V., and M.L.-B.; methodology, A.C., P.S., and O.T.-P.; software, A.C.; validation, A.C. and O.T.-P.; formal analysis, A.C.; investigation, A.C. and P.S.; resources, M.L.-B. and D.V.; data curation, A.C.; writing—original draft preparation, A.C.; writing—review and editing, A.C., P.S., O.T.-P., M.L.-B., and D.V.; visualization, O.T.-P. and M.L.-B.; supervision, M.L.-B., C.E., M.I., and D.V.; project administration, D.V.; funding acquisition, C.E., M.I., and D.V.

Funding: This research was financially supported by Aigües de Barcelona and sponsored by Diputació de Barcelona (DiBa) in the context of the Biobesòs Project (Evaluation of the wetlands' efficacy to remove pollutants and improve the biodiversity levels in the Besòs river and its surroundings).

Acknowledgments: The authors thank Pau Fortuño, Narcís Prat, Núria Flor, and Núria Cid from Universitat de Barcelona (Barcelona, Spain) for their excellent help with sampling and technical assistance. We are grateful to Juan Carlos Ruiz and María Monzó from Aigües de Barcelona, and Anna Riera and Antoni Maza from DiBa for the support to make this project possible.

Conflicts of Interest: The authors declare no conflicts of interest.

References

1. Pankhurst, N.W. The endocrinology of stress in fish: An environmental perspective. *Gen. Comp. Endocrinol.* **2011**, *170*, 265–275. [CrossRef] [PubMed]
2. Schreck, C.B.; Tort, L. *Biology of Stress in Fish*, 2016th ed.; Farrell, A.P., Brauner, C.J., Eds.; Academic Press: Cambridge, MA, USA, 2016; ISBN 9780128027288.
3. Sheriff, M.J.; Dantzer, B.; Delehanty, B.; Palme, R.; Boonstra, R. Measuring stress in wildlife: Techniques for quantifying glucocorticoids. *Oecologia* **2011**, *166*, 869–887. [CrossRef] [PubMed]
4. Wingfield, J.C.; Kelley, J.P.; Angelier, F. What are extreme environmental conditions and how do organisms cope with them? *Curr. Zool.* **2011**, *57*, 363–374. [CrossRef]
5. Scott, G.R.; Sloman, K.A. The effects of environmental pollutants on complex fish behaviour: Integrating behavioural and physiological indicators of toxicity. *Aquat. Toxicol.* **2004**, *68*, 369–392. [CrossRef]
6. Dantzer, B.; Fletcher, Q.E.; Boonstra, R.; Sheriff, M.J. Measures of physiological stress: A transparent or opaque window into the status, management and conservation of species? *Conserv. Physiol.* **2014**, *2*, 1–18. [CrossRef]
7. Hooper, D.U.; Adair, E.C.; Cardinale, B.J.; Byrnes, J.E.K.; Hungate, B.A.; Matulich, K.L.; Gonzalez, A.; Duffy, J.E.; Gamfeldt, L.; Connor, M.I. A global synthesis reveals biodiversity loss as a major driver of ecosystem change. *Nature* **2012**, *486*, 105–108. [CrossRef]
8. Jeffrey, J.D.; Hasler, C.T.; Chapman, J.M.; Cooke, S.J.; Suski, C.D. Linking landscape-scale disturbances to stress and condition of fish: Implications for restoration and conservation. *Integr. Comp. Biol.* **2015**, *55*, 618–630. [CrossRef]
9. Johnstone, C.P.; Reina, R.D.; Lill, A. Interpreting indices of physiological stress in free-living vertebrates. *J. Comp. Physiol. B* **2012**, *182*, 861–879. [CrossRef]
10. Maceda-Veiga, A.; Figuerola, J.; Martínez-Silvestre, A.; Viscor, G.; Ferrari, N.; Pacheco, M. Inside the Redbox: Applications of haematology in wildlife monitoring and ecosystem health assessment. *Sci. Total Environ.* **2015**, *514*, 322–332. [CrossRef]
11. Colin, N.; Maceda-Veiga, A.; Monroy, M.; Ortega-Ribera, M.; Llorente, M.; de Sostoa, A. Trends in biomarkers, biotic indices, and fish population size revealed contrasting long-term effects of recycled water on the ecological status of a Mediterranean river. *Ecotoxicol. Environ. Saf.* **2017**, *145*, 340–348. [CrossRef]
12. Hussain, B.; Sultana, T.; Sultana, S.; Masoud, M.S.; Ahmed, Z.; Mahboob, S. Fish eco-genotoxicology: Comet and micronucleus assay in fish erythrocytes as in situ biomarker of freshwater pollution. *Saudi J. Biol. Sci.* **2018**, *25*, 393–398. [CrossRef] [PubMed]
13. Braham, R.P.; Blazer, V.S.; Shaw, C.H.; Mazik, P.M. Micronuclei and other erythrocyte nuclear abnormalities in fishes from the Great Lakes Basin, USA. *Environ. Mol. Mutagen.* **2017**, *58*, 570–581. [CrossRef] [PubMed]
14. Castaño, A.; Sanchez, P.; Llorente, M.T.; Carballo, M.; De La Torre, A.; Muñoz, M.J. The use of alternative systems for the ecotoxicological screening of complex mixtures on fish populations. *Sci. Total Environ.* **2000**, *247*, 337–348. [CrossRef]
15. Ivanova, L.; Popovska-percinic, F.; Slavevska-stamenkovic, V.; Jordanova, M.; Rebok, K. Micronuclei and nuclear abnormalities in erythrocytes from barbel barbus peloponnesius revealing genotoxic pollution of the River Bregalnica. *Maced. Vet. Rev.* **2016**, *39*, 159–166. [CrossRef]
16. Davis, A.K.; Maney, D.L.; Maerz, J.C. The use of leukocyte profiles to measure stress in vertebrates: A review for ecologists. *Funct. Ecol.* **2008**, *22*, 760–772. [CrossRef]
17. Hedayati, A.; Jahanbakhshi, A. The effect of water-soluble fraction of diesel oil on some hematological indices in the great sturgeon Huso huso. *Fish Physiol. Biochem.* **2012**, *38*, 1753–1758. [CrossRef]
18. Witeska, M. Stress in fish-hematological and immunological effects of heavy metals. *Electron. J. Ichthyol.* **2005**, *1*, 35–41.
19. Busch, D.S.; Hayward, L.S. Stress in a conservation context: A discussion of glucocorticoid actions and how levels change with conservation-relevant variables. *Biol. Conserv.* **2009**, *142*, 2844–2853. [CrossRef]
20. Homyack, J.A. Evaluating habitat quality of vertebrates using conservation physiology tools. *Wildl. Res.* **2010**, *37*, 332–342. [CrossRef]

21. Mommsen, T.P.; Vijayan, M.M.; Moon, T.W. Cortisol in teleosts: Dynamics, mechanisms of action, and metabolic regulation. *Rev. Fish Biol. Fish.* **1999**, *9*, 211–268. [CrossRef]
22. King, G.D.; Chapman, J.M.; Cooke, S.J.; Suski, C.D. Stress in the neighborhood: Tissue glucocorticoids relative to stream quality for five species of fish. *Sci. Total Environ.* **2016**, *547*, 87–94. [CrossRef] [PubMed]
23. Norris, D.O.; Donahue, S.; Dores, R.M.; Lee, J.K.; Tammy, A.; Ruth, T.; Woodling, J.D. Impaired Adrenocortical Response to Stress by Brown Trout, Salmo trutta, Living in Metal-Contaminated Waters of the Eagle River, Colorado. *Gen. Comp. Endocrinol.* **1999**, *113*, 1–8. [CrossRef] [PubMed]
24. Pottinger, T.G.; Williams, R.J.; Matthiessen, P. A comparison of two methods for the assessment of stress axis activity in wild fish in relation to wastewater effluent exposure. *Gen. Comp. Endocrinol.* **2016**, *230*, 29–37. [CrossRef] [PubMed]
25. Aerts, J.; Metz, J.R.; Ampe, B.; Decostere, A.; Flik, G.; De Saeger, S. Scales Tell a Story on the Stress History of Fish. *PLoS ONE* **2015**, *10*, 0123411. [CrossRef]
26. Carbajal, A.; Reyes-López, F.E.; Tallo-Parra, O.; Lopez-Bejar, M.; Tort, L. Comparative assessment of cortisol in plasma, skin mucus and scales as a measure of the hypothalamic-pituitary-interrenal axis activity in fish. *Aquaculture* **2019**, *506*, 410–416. [CrossRef]
27. Carbajal, A.; Tallo-Parra, O.; Monclús, L.; Vinyoles, D.; Solé, M.; Lacorte, S.; Lopez-Bejar, M. Variation in scale cortisol concentrations of a wild freshwater fish: Habitat quality or seasonal influences? *Gen. Comp. Endocrinol.* **2019**, *275*, 44–50. [CrossRef]
28. De Mercado, E.; Larrán, A.M.; Pinedo, J.; Tomás-Almenar, C. Skin mucous: A new approach to assess stress in rainbow trout. *Aquaculture* **2018**, *484*, 90–97. [CrossRef]
29. Guardiola, F.A.; Cuesta, A.; Esteban, M.Á. Using skin mucus to evaluate stress in gilthead seabream (*Sparus aurata* L.). *Fish Shellfish Immunol.* **2016**, *59*, 323–330. [CrossRef]
30. Simontacchi, C.; Poltronieri, C.; Carraro, C.; Bertotto, D.; Xiccato, G.; Trocino, A.; Radaelli, G. Alternative stress indicators in sea bass *Dicentrarchus labrax*, L. *J. Fish Biol.* **2008**, *72*, 747–752. [CrossRef]
31. Bertotto, D.; Poltronieri, C.; Negrato, E.; Majolini, D.; Radaelli, G.; Simontacchi, C. Alternative matrices for cortisol measurement in fish. *Aquac. Res.* **2010**, *41*, 1261–1267. [CrossRef]
32. Fernández-Alacid, L.; Sanahuja, I.; Ordóñez-Grande, B.; Sánchez-Nuño, S.; Herrera, M.; Ibarz, A. Skin mucus metabolites and cortisol in meagre fed acute stress-attenuating diets: Correlations between plasma and mucus. *Aquaculture* **2019**, *499*, 185–194. [CrossRef]
33. Franco-Martinez, L.; Tvarijonaviciute, A.; Martinez-Subiela, S.; Teles, M.; Tort, L. Chemiluminescent assay as an alternative to radioimmunoassay for the measurement of cortisol in plasma and skin mucus of Oncorhynchus mykiss. *Ecol. Indic.* **2019**, *98*, 634–640. [CrossRef]
34. Doadrio, I.; Kottelat, M.; de Sostoa, A. Squalius laietanus, a new species of cyprinid fish from north-eastern Spain and southern France (*Teleostei: Cyprinidae*). *Ichthyol. Explor. Freshw.* **2007**, *18*, 247–256.
35. Babatunde, A.O.; Zhao, Y.Q.; O'Neill, M.; O'Sullivan, B. Constructed wetlands for environmental pollution control: A review of developments, research and practice in Ireland. *Environ. Int.* **2008**, *34*, 116–126. [CrossRef]
36. Brix, H.; Schierup, H.; Bourke, J.; Wroe, S.; Moreno, K.; McHenry, C.; Clausen, P.; Aller, L.; Lehr, J.; Petty, R.; et al. Use of Constructed Wetlands in Water Pollution Control: Historical Development, Present Status, and Future Perspectives. *PLoS ONE* **1994**, *3*, 209–223. [CrossRef]
37. Vymazal, J.; Březinová, T. The use of constructed wetlands for removal of pesticides from agricultural runoff and drainage: A review. *Environ. Int.* **2015**, *75*, 11–20. [CrossRef]
38. Wu, H.; Zhang, J.; Ngo, H.H.; Guo, W.; Hu, Z.; Liang, S.; Fan, J.; Liu, H. A review on the sustainability of constructed wetlands for wastewater treatment: Design and operation. *Bioresour. Technol.* **2015**, *175*, 594–601. [CrossRef]
39. Guittonny-Philippe, A.; Masotti, V.; Höhener, P.; Boudenne, J.L.; Viglione, J.; Laffont-Schwob, I. Constructed wetlands to reduce metal pollution from industrial catchments in aquatic Mediterranean ecosystems: A review to overcome obstacles and suggest potential solutions. *Environ. Int.* **2014**, *64*, 1–16. [CrossRef]
40. Gatz, A.J.; Linder, R.S. Effects of Repeated Electroshocking on Condition, Growth, and Movement of Selected Warmwater Stream Fishes. *N. Am. J. Fish. Manag.* **2008**, *28*, 792–798. [CrossRef]
41. Nagrodski, A.; Murchie, K.J.; Stamplecoskie, K.M.; Suski, C.D.; Cooke, S.J. Effects of an experimental short-term cortisol challenge on the behaviour of wild creek chub Semotilus atromaculatus in mesocosm and stream environments. *J. Fish Biol.* **2013**, *82*, 1138–1158. [CrossRef]

42. Maceda-Veiga, A.; Monroy, M.; Navarro, E.; Viscor, G.; de Sostoa, A. Metal concentrations and pathological responses of wild native fish exposed to sewage discharge in a Mediterranean river. *Sci. Total Environ.* **2013**, *449*, 9–19. [CrossRef] [PubMed]
43. Stasinakis, A.S.; Mermigka, S.; Samaras, V.G.; Farmaki, E.; Thomaidis, N.S. Occurrence of endocrine disrupters and selected pharmaceuticals in Aisonas River (Greece) and environmental risk assessment using hazard indexes. *Environ. Sci. Pollut. Res.* **2012**, *19*, 1574–1583. [CrossRef] [PubMed]
44. Romero, L.M. Physiological stress in ecology: Lessons from biomedical research. *Trends Ecol. Evol.* **2004**, *19*, 249–255. [CrossRef] [PubMed]
45. Schultz, D.R.; Perez, N.; Tan, C.K.; Mendez, A.J.; Capo, T.R.; Snodgrass, D.; Prince, E.D.; Serafy, J.E. Concurrent levels of 11-ketotestosterone in fish surface mucus, muscle tissue and blood. *J. Appl. Ichthyol.* **2005**, *21*, 394–398. [CrossRef]
46. Goodbred, S.L.; Patiño, R.; Torres, L.; Echols, K.R.; Jenkins, J.A.; Rosen, M.R.; Orsak, E. Are endocrine and reproductive biomarkers altered in contaminant-exposed wild male Largemouth Bass (*Micropterus salmoides*) of Lake Mead, Nevada/Arizona, USA? *Gen. Comp. Endocrinol.* **2015**, *219*, 125–135. [CrossRef] [PubMed]
47. Barton, B.A.; Morgan, J.D.; Vljayan, M. Physiological and Condition-Related Indicators of Environmental Stress in Fish. *Biol. Indic. Aquat. Ecosyst. Stress* **1998**, 111–148.
48. Pacheco, M.; Santos, M.A. Induction of Micronuclei ad nuclear abnormalities in the erythrocytes of Anguilla angulila L. exposed either to cyclophosphamide or to bleached kraft pulp mill effluent. *Fresenius Environ. Bull.* **1996**, *5*, 746–751.
49. Tavares-Dias, M. Cytochemical method for staining fish basophils. *J. Fish Biol.* **2006**, *69*, 312–317. [CrossRef]
50. Carbajal, A.; Monclús, L.; Tallo-Parra, O.; Sabes-Alsina, M.; Vinyoles, D.; Lopez-Bejar, M. Cortisol detection in fish scales by enzyme immunoassay: Biochemical and methodological validation. *J. Appl. Ichthyol.* **2018**, *34*, 967–970. [CrossRef]
51. Maceda-Veiga, A.; Monroy, M.; Viscor, G.; De Sostoa, A. Changes in non-specific biomarkers in the Mediterranean barbel (*Barbus meridionalis*) exposed to sewage effluents in a Mediterranean stream (Catalonia, NE Spain). *Aquat. Toxicol.* **2010**, *100*, 229–237. [CrossRef]
52. Colin, N.; Porte, C.; Fernandes, D.; Barata, C.; Padrós, F.; Carrassón, M.; Monroy, M.; Cano-Rocabayera, O.; de Sostoa, A.; Piña, B.; et al. Ecological relevance of biomarkers in monitoring studies of macro-invertebrates and fish in Mediterranean rivers. *Sci. Total Environ.* **2016**, *540*, 307–323. [CrossRef] [PubMed]
53. Pacheco, M.; Santos, M.A. Biotransformation, genotoxic, and histopathological effects of environmental contaminants in European eel (*Anguilla anguilla* L.). *Ecotox. Environ. Saf.* **2002**, *53*, 331–347. [CrossRef]
54. Mussali-Galante, P.; Tovar-Sánchez, E.; Valverde, M.; Rojas del Castillo, E. Biomarkers of Exposure for Assessing Environmental Metal Pollution: From Molecules To Ecosystems. *Rev. Int. Contam. Ambient.* **2013**, *29*, 117–140.
55. Botelho, R.G.; Monteiro, S.H.; Christofoletti, C.A.; Moura-Andrade, G.C.R.; Tornisielo, V.L. Environmentally Relevant Concentrations of Atrazine and Ametrine Induce Micronuclei Formation and Nuclear Abnormalities in Erythrocytes of Fish. *Arch. Environ. Contam. Toxicol.* **2015**, *69*, 577–585. [CrossRef]
56. Ahmad, I.; Ahmad, M. FreshWater Fish, Channa Punctatus, as a Model for Pendimethalin Genotoxicity Testing: A New Approach toward Aquatic Environmental Contaminants. *Environ. Toxicol.* **2016**, *31*, 1520–1529. [CrossRef]
57. Spencer, H.B.; Hussein, W.R.; Tchounwou, P.B. Effects of tetrachloroethylene on the viability and development of embryos of the Japanese Medaka, Oryzias latipes. *Arch. Environ. Contam. Toxicol.* **2002**, *42*, 463–469. [CrossRef]
58. Wang, G.; Zhou, Q.; Hu, X.; Hua, T.; Li, F. Single and joint toxicity of perchloroethylene and cadmium on Ctenopharyngodon idellus. *J. Appl. Ecol.* **2007**, *18*, 1120–1124.
59. Augusto, L.G.D.S.; Lieber, S.R.; Ruiz, M.A.; De Souza, C.A. Micronucleus monitoring to assess human occupational exposure to organochlorides. *Environ. Mol. Mutagen.* **1997**, *29*, 46–52. [CrossRef]
60. Gil Barcellos, L.J.; Kreutz, L.C.; de Souza, C.; Rodrigues, L.B.; Fioreze, I.; Quevedo, R.M.; Cericato, L.; Soso, A.B.; Fagundes, M.; Conrad, J.; et al. Hematological changes in jundiá (*Rhamdia quelen* Quoy and Gaimard Pimelodidae) after acute and chronic stress caused by usual aquacultural management, with emphasis on immunosuppressive effects. *Aquaculture* **2004**, *237*, 229–236. [CrossRef]
61. Grzelak, A.K.; Davis, D.J.; Caraker, S.M.; Crim, M.J.; Spitsbergen, J.M.; Wiedmeyer, C.E. Stress leukogram induced by acute and chronic stress in zebrafish (*Danio rerio*). *Comp. Med.* **2017**, *67*, 263–269.

62. Wendelaar Bonga, S.E. The Stress Response in Fish. *Physiol. Rev.* **1997**, *77*, 591–625. [CrossRef] [PubMed]
63. Letcher, R.J.; Ove, J.; Dietz, R.; Jenssen, B.M.; Jørgensen, E.H.; Sonne, C.; Verreault, J.; Vijayan, M.M.; Gabrielsen, G.W. Exposure and effects assessment of persistent organohalogen contaminants in arctic wildlife and fish. *Sci. Total Environ.* **2010**, *408*, 2995–3043. [CrossRef] [PubMed]
64. Tavares-Dias, M. A morphological and cytochemical study of erythrocytes, thrombocytes and leukocytes in four freshwater teleosts. *J. Fish Biol.* **2006**, *68*, 1822–1833. [CrossRef]
65. Hedayati, A.; Hassan Nataj Niazie, E. Hematological changes of silver carp (*Hypophthalmichthys molitrix*) in response to Diazinon pesticide. *J. Environ. Heal. Sci. Eng.* **2015**, *13*, 2–6. [CrossRef] [PubMed]
66. Cook, N.J. Review: Minimally invasive sampling media and the measurement of corticosteroids as biomarkers of stress in animals. *Can. J. Anim. Sci.* **2012**, *92*, 227–259. [CrossRef]
67. Knag, A.C.; Taugbøl, A. Acute exposure to offshore produced water has an effect on stress- and secondary stress responses in three-spined stickleback Gasterosteus aculeatus. *Comp. Biochem. Physiol. C Toxicol. Pharmacol.* **2013**, *158*, 173–180. [CrossRef]
68. Pottinger, T.G.; Matthiessen, P. Disruption of the stress response in wastewater treatment works effluent-exposed three-spined sticklebacks persists after translocation to an unpolluted environment. *Ecotoxicology* **2016**, *25*, 538–547. [CrossRef]
69. Baos, R.; Blas, J.; Bortolotti, G.R.; Marchant, T.A.; Hiraldo, F. Adrenocortical response to stress and thyroid hormone status in free-living nestling white storks (*Ciconia ciconia*) exposed to heavy metal and arsenic contamination. *Environ. Health Perspect.* **2006**, *114*, 1497–1501. [CrossRef]
70. Strong, R.J.; Pereira, M.G.; Shore, R.F.; Henrys, P.A.; Pottinger, T.G. Feather corticosterone content in predatory birds in relation to body condition and hepatic metal concentration. *Gen. Comp. Endocrinol.* **2015**, *214*, 47–55. [CrossRef]
71. Wikelski, M.; Wong, V.; Chevalier, B.; Rattenborg, N.; Snell, H.L. Marine iguanas die from trace oil pollution. *Nature* **2002**, *417*, 607–608. [CrossRef]
72. Blevins, Z.W.; Wahl, D.H.; Suski, C.D. Reach-Scale Land Use Drives the Stress Responses of a Resident Stream Fish. *Physiol. Biochem. Zool.* **2013**, *87*, 113–124. [CrossRef] [PubMed]
73. Gesto, M.; Soengas, J.L.; Míguez, J.M. Acute and prolonged stress responses of brain monoaminergic activity and plasma cortisol levels in rainbow trout are modified by PAHs (naphthalene, β-naphthoflavone and benzo(a)pyrene) treatment. *Aquat. Toxicol.* **2008**, *86*, 341–351. [CrossRef] [PubMed]
74. Oliveira, M.; Pacheco, M.; Santos, M.A. Fish thyroidal and stress responses in contamination monitoring-An integrated biomarker approach. *Ecotoxicol. Environ. Saf.* **2011**, *74*, 1265–1270. [CrossRef] [PubMed]
75. Pottinger, T.G.; Henrys, P.A.; Williams, R.J.; Matthiessen, P. The stress response of three-spined sticklebacks is modified in proportion to effluent exposure downstream of wastewater treatment works. *Aquat. Toxicol.* **2013**, *126*, 382–392. [CrossRef]
76. Boonstra, R. The ecology of stress: A marriage of disciplines. *Funct. Ecol.* **2013**, *27*, 7–10. [CrossRef]
77. Baker, M.R.; Gobush, K.S.; Vynne, C.H. Review of factors influencing stress hormones in fish and wildlife. *J. Nat. Conserv.* **2013**, *21*, 309–318. [CrossRef]
78. Houston, A.H. Review: Are the Classical Hematological Variables Acceptable Indicators of Fish Health? *Trans. Am. Fish. Soc.* **1997**, *126*, 879–894. [CrossRef]
79. Farag, M.R.; Alagawany, M. Erythrocytes as a biological model for screening of xenobiotics toxicity. *Chem. Biol. Interact.* **2018**, *279*, 73–83. [CrossRef]
80. Nikinmaa, M. How does environmental pollution affect red cell function in fish? *Aquat. Toxicol.* **1992**, *22*, 227–238. [CrossRef]
81. Dickens, M.J.; Romero, L.M. A consensus endocrine profile for chronically stressed wild animals does not exist. *Gen. Comp. Endocrinol.* **2013**, *191*, 177–189. [CrossRef]

MDPI
St. Alban-Anlage 66
4052 Basel
Switzerland
Tel. +41 61 683 77 34
Fax +41 61 302 89 18
www.mdpi.com

Animals Editorial Office
E-mail: animals@mdpi.com
www.mdpi.com/journal/animals

www.ingramcontent.com/pod-product-compliance
Lightning Source LLC
LaVergne TN
LVHW070925170726
843515LV00005B/872

* 9 7 8 3 0 3 6 5 0 1 4 2 0 *